JDN 24 - Neutrons and Biology

EPJ Web of Conferences Volume 236 (2018)

Carqueiranne, France
16 - 19 September 2018

Editors:

Sophie Combet
Giorgio Schirò

ISBN: 978-1-7138-1441-2

Printed from e-media with permission by:

Curran Associates, Inc.
57 Morehouse Lane
Red Hook, NY 12571

Printed with permission by Curran Associates, Inc. (2020)

For additional information, please contact EDP Sciences – Web of Conferences at the address below.

EDP Sciences – Web of Conferences
17, Avenue du Hoggar
Parc d'Activité de Courtabœuf
BP 112
F-91944 Les Ulis Cedex A
France

Phone: +33 (0) 1 69 18 75 75
Fax: +33 (0) 1 69 28 84 91

contact-edps@webofconferences.org

Additional copies of this publication are available from:

Curran Associates, Inc.
57 Morehouse Lane
Red Hook, NY 12571 USA
Phone: 845-758-0400
Fax: 845-758-2633
Email: curran@proceedings.com
Web: www.proceedings.com

TABLE OF CONTENTS

AN INTRODUCTION TO NEUTRONS FOR BIOLOGY 1
Combet Sophie

OPPORTUNITIES AND CHALLENGES IN NEUTRON CRYSTALLOGRAPHY 19
Zaccai Nathan Richard, Coquelle Nicolas

THE BASICS OF SMALL-ANGLE NEUTRON SCATTERING (SANS FOR NEW USERS OF STRUCTURAL BIOLOGY) 41
Jeffries Cy M., Pietras Zuzanna, Svergun Dmitri I.

THE POWER OF SANS, COMBINED WITH DEUTERATION AND CONTRAST VARIATION, FOR STRUCTURAL STUDIES OF FUNCTIONAL AND DYNAMIC BIOMACROMOLECULAR SYSTEMS IN SOLUTION 76
Mahieu Emilie, Ibrahim Ziad, Moulin Martine, Härtlein Michael, Franzetti Bruno, Martel Anne, Gabel Frank

SOLUTION STRUCTURE OF MACROMOLECULES USING SMALL ANGLE NEUTRON SCATTERING AND MOLECULAR SIMULATIONS 86
Bhatt Jayesh S.

AN INTRODUCTION TO NEUTRON REFLECTOMETRY 101
Cousin Fabrice, Fadda Giulia

APPLICATIONS OF NEUTRON REFLECTOMETRY IN BIOLOGY 116
Gerelli Yuri

PROBING THE DYNAMICS OF BIOLOGICAL MATTER BY ELASTIC, QUASI-ELASTIC, AND INELASTIC NEUTRON SCATTERING 134
Schirò Giorgio

Author Index

Foreword

The 2018 edition of the "Ecole thématique de la Société Française de la Neutronique (SFN)" was organized as a four-day workshop in Carqueiranne, France, from September 16th to 19th, and focused on **"Neutrons & Biology"**. This school was organized with the support of CNRS (formation permanente), Laboratoire Léon-Brillouin (LLB, France), Institut Laue-Langevin (ILL, France), Fédération Française de Diffusion Neutronique (2FDN), Institut de Biologie Structurale (IBS, France), and Université Grenoble-Alpes (UGA, France).

The school developed the following thematic sessions:

- a **general introduction** to neutron techniques (José Teixeira, LLB),
- a session dedicated to **structural studies** in biology, in particular:
  - **neutron crystallography**, composed by an introduction to the theory of protein crystallography (Nathan Zaccai, Univ. Cambridge, UK) and a lecture on instrumentation (Nicolas Coquelle, ILL),
  - **small-angle neutron scattering (SANS)**, with a theoretical introduction (Cy Jeffries, EMBL, Germany) and a lecture on SANS instrumentation (Anne Martel, ILL),
  - **neutron reflectometry** theory and instrumentation (Fabrice Cousin, LLB),
- a session dedicated to the use of **inelastic neutron scattering** for studying biomolecular **dynamics** composed by an introduction to theory (Giorgio Schirò, IBS) and a lecture on instrumentation (Andrea Orecchini, Univ. Perugia, Italy),
- three lectures on the integrated use of neutrons and complementary techniques in biology (**transversal studies**) by Nathan Zaccai, Frank Gabel (IBS), and Jayesh Bhatt (Univ. College London, UK),
- a large part of the school was also dedicated to **tutorials** on neutron crystallography (Nicolas Coquelle), SANS (Sophie Combet, LLB, and Anne Martel), quasi-elastic and inelastic neutron scattering (Andrea Orecchini), and reflectometry (Sebastian Koehler, ILL),
- and, last but not least, a session to learn **how to write a proposal** to succeed in obtaining neutron beam time (Giulia Fadda, LLB).

This issue is mostly dedicated to the lectures of the school.

Enjoy the reading!

The organizing committee: Sophie Combet, Viviana Cristiglio (ILL), Giulia Fadda, Anne Martel, and Giorgio Schirò.

edp sciences

EPJ Web of Conferences

Statement of Peer review

In submitting conference proceedings to EPJ Web of Conferences, I certify to the Publisher that I adhere to the **Policy on Publishing Integrity** of the journal in order to safeguard good scientific practice in publishing.

1. All articles have been subjected to peer review administered by the proceedings editors.
2. Reviews have been conducted by expert referees, who have been requested to provide unbiased and constructive comments aimed, whenever possible, at improving the work.
3. Proceedings editors have taken all reasonable steps to ensure the quality of the materials they publish and their decision to accept or reject a paper for publication has been based only on the merits of the work and the relevance to the journal.

Title, date and place of the conference

Neutrons & Biology
September 16-19, 2018
Carqueiranne, France

Proceedings editor(s):

Sophie Combet
Giorgio Schirò

Date and editor's signature

June 11, 2020

EPJ Web of Conferences **236**, 01001 (2020) https://doi.org/10.1051/epjconf/202023601001
JDN 24

An introduction to neutrons for biology

Sophie Combet[1,*]
[1]Laboratoire Léon-Brillouin (LLB), UMR 12 CEA-CNRS, Université Paris-Saclay, CEA-Saclay, 91191 Gif-sur-Yvette CEDEX, France

Abstract. The overlap of biology and neutron scattering remains a relatively narrow domain of research. This is partly due to the *a priori* maladjustment between real space problems and methods based on spatial and temporal correlations. In addition, some major assets of neutron scattering, such as isotopic substitution, can be tricky with biological molecules. More generally, a mutual lack of knowledge of the two concerned communities precluded potential rich interactions in early times. However, the situation changed to the point that, today, biology represents a substantial part of the research activity at neutron facilities. The purpose of this introduction is not to present one more overview of the subject of "neutron scattering" (excellent comprehensive articles are easily accessible to the interested readers [1–4]), but rather to facilitate the reading of the present book by introducing a few neutron scattering notions that may be useful for the community of biologists eventually less familiar with this technique.

1 How can neutrons be a useful probe in biology?

Biology is a wide field whose subjects of study are as diverse as populations, organisms, cells, or macromolecules. Actually, almost only biologists interested by biological macromolecules (proteins, nucleic acids, or lipid membranes constituting the living cells) or by issues at the molecular scale can be concerned by neutron scattering techniques.

A crucial issue in biology remains the knowledge of how atoms and molecules organize themselves and move to accomplish a biological function. The long-term "dream" is to observe macromolecules at work *in vivo* and to describe not only the positions of atoms but also the atomic motions and conformational changes that occur.

To probe structure, interactions, or dynamics of biomacromolecules, the "classical" techniques currently used by biologists are X-ray crystallography, nuclear magnetic resonance (NMR) spectroscopy, or the increasingly used cryo-electron microscopy. However, neutron scattering, as a complementary or even alternative technique, can prove to be essential to answer specific but relevant questions of biology.

In practice, the success of any technical approach depends on its insertion in modern biology through several parameters: high throughput, automatization, facilities to label the samples, perform the experiment, treat the data, etc. Today, neutron scattering is still "under-used" in biology, where it occupies a kind of "niche". Such situation reduces its potential

* Corresponding author: sophie.combet@cea.fr

importance and impact in the community of biologists. This is partly due to the *a priori* maladjustment between real space problems in biology and scattering methods, which are based on spatial and temporal correlations. More deeply, this under-utilization may be due to communication difficulties between physicists and biologists. For instance, the notion of "complexity" in biology drastically prevents or makes very difficult the practice of mathematical tools always present in physical models. There is therefore a "risk" for neutron users to move from biology issues to macromolecular physics.

This situation is also due to more "practical" reasons, such as difficulties of access, since neutron beams are produced only at nuclear reactors or spallation sources and thus are not accessible on a laboratory bench. The quantity and quality of sample necessary for a neutron experiment can sometimes be an obstacle in biology, although it depends on the beam intensity and the instrument used [5]. A special attention has indeed to be paid to the quality control of sample purity, monodispersity, and the monomeric state of the sample (*e.g.* by using SDS-PAGE, analytical NMR, SEC-DLS, or SEC-MALS) before but also all along an experiment, whose duration (from a few minutes to several days) can be tricky for most of biomolecules [5] (see also the book section written by Jeffries *et al.*). As such, the second generation of neutron spallation sources, like the European Spallation Source (ESS, Lund, Sweden) that should be operational in the next years, will open up new perspectives.

Like other neutron users, biologists can be driven by local contacts to perform a neutron experiment. As mentioned above, access to neutron experiments remains difficult since beam time is mainly accessible through experiment proposals evaluated by *ad hoc* scientific committees. Besides, there are many different home-made methods to reduce the data in neutron facilities that can make data treatment difficult but local contacts are usually available to provide help for this step.

Fortunately, recently, the rapid and huge development of neutron instruments, sample environments, tools for data treatment, and computer simulations [6,7] changed the situation to the point that nowadays biology represents a substantial part of the research activity at neutron facilities.

2 The neutron scattering technique

2.1 Properties of neutrons

Neutrons are nuclear particles discovered by J. Chadwick in 1932 (less than one century ago!) and were used the first time in 1946 by C. Shull as a tool for scattering experiments. Neutrons can be produced by two different ways (nuclear fission or spallation). Neutron scattering technique has since been developed to concern now all the aspects of condensed matter: physics, chemistry, materials, and so, as far as we are concerned, biology. Neutrons are an exceptional tool because they interact with atomic nuclei with very small energy, therefore without physical or chemical damage of the samples.

The energy of thermal neutrons is, by definition, that of room temperature, *i.e.* ~25 meV, since they are thermalized at room temperature. Such domain of energy is often that of molecular motions (diffusion, rotation) and of chemical reactions. From this point of view, biology is an ideal domain to apply neutron scattering, as many biological processes take place also at room temperature. However, the large size of biomolecules, even of their secondary and tertiary structures, implies motions too slow to be accessible to neutrons, limiting substantially the potential dynamical studies.

Indeed, with typical wavelengths between 1 and 10 Å, neutrons cover ideally the nm and the ps space-time domain. Modern instruments succeed to extend this domain to two or three orders of magnitude, in the domain of hundreds of nm or several ns, but with a lower quality of the data and longer data acquisition times.

2.2 Principles of neutron scattering

Quantum mechanics associates any moving particle to a wave of wavelength $\lambda = \frac{h}{p}$ and energy $E = \hbar\omega$, where $h = 2\pi\hbar$ is the Planck constant and $p = mv$ is the momentum of a particle of mass m moving with the velocity v.

Like in other scattering experiments, a neutron scattering experiment consists of sending a monochromatic beam of particles to a sample. Incident neutrons are defined by the momentum, thus by the wave vector $\vec{k}_i$, and energy E_i:

$$\|\vec{k}_i\| = \frac{2\pi}{\lambda_i}$$

$$E_i = \frac{\hbar^2 k_i^2}{2m} = \left(\frac{h^2}{2m}\right)\left(\frac{1}{\lambda_i^2}\right) = \frac{1}{2} m v_i^2$$

Neutrons interact with the nuclei of the sample through the nuclear interaction potential. They are scattered by the sample and measured in function of their wave vector $\vec{k}_s$ and energy E_s, providing information about the properties of the sample.

The scattering vector is defined by the wave vector $\vec{Q}$ (Fig. 1) and the exchanged energy E:

$$\vec{Q} = \vec{k}_i - \vec{k}_s$$

$$E = \hbar\omega = E_i - E_s$$

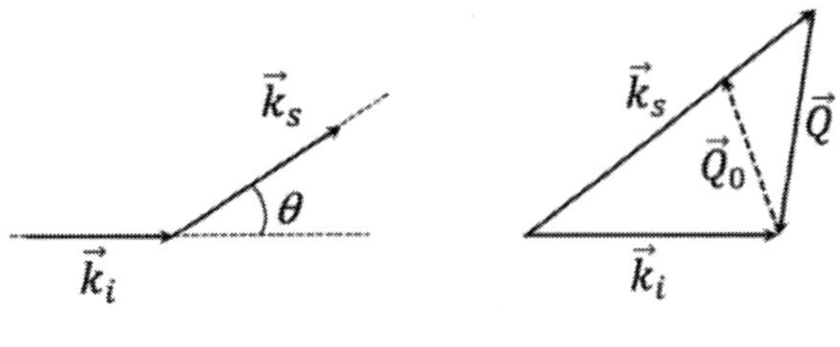

Fig. 1. Principle of a scattering experiment in the reciprocal space. Due to interaction with atoms of the sample, neutrons undergo a modification of their wave vector and energy. The case $\vec{Q} = \vec{Q}_0$ corresponds to elastic scattering, for which $\|\vec{k}_i\| = \|\vec{k}_s\|$ and $\omega = 0$.

In the case of inelastic scattering ($\|\vec{k}_i\| \neq \|\vec{k}_s\|$), we can define the modulus of the vector $\vec{Q}$ (Fig. 1) by:

$$Q = \sqrt{k_i^2 + k_s^2 - 2k_i k_s cos\theta} \quad \textbf{Eq. 1}$$

In elastic scattering, *i.e.* without any energy exchange between the neutrons and the sample ($E_i = E_s, \|\vec{k}_i\| = \|\vec{k}_s\|$), we can define the modulus of the elastic scattering vector, noted $\vec{Q}_0$ (Fig. 1) by:

$$Q_0 = \frac{4\pi}{\lambda} \sin{(\theta/2)}$$

where θ is the scattering angle. Note that Q depends on the instrumental conditions (λ, θ), not on the sample.

Each incident neutron is described by a planar wave defined by a wave vector $\vec{k}$ that, when interacting with a single atom, is scattered isotropically in all directions, generating a spherical wave because the wavelength (10^{-8} cm) is much larger than the size of an atomic nucleus (of the order of fm = 10^{-13} cm) (Fig. 2). The anisotropic nature of the scattering is due to the interference of multiple spherical waves centered in the different nuclei that constitute the sample.

Note that neutron scattering does not measure distances like a microscope would do. It measures correlation lengths and correlation times in the reciprocal space.

2.3 Interferences

Since the atomic nuclei are much smaller than the wavelengths used for neutron scattering, the nucleus acts as a point scatterer, resulting in isotropic scattering. For an incident plane wave described by the wave function Ψ, with unit amplitude, travelling in the positive z direction with wavenumber $k_i = \|\vec{k}_i\|$, the scattered wave Ψ' is then spherical (Fig. 2), with an amplitude $\frac{b}{r}$ specific of the atomic nucleus, where r is the distance from the nucleus (placed at $r = 0$):

$$\Psi = e^{ik_i z}$$

$$\Psi' = -\frac{b}{r} e^{ik_s r}$$

The positive constant b is the scattering length of a nucleus and is a measure of the "strength" of its interaction with the neutrons.

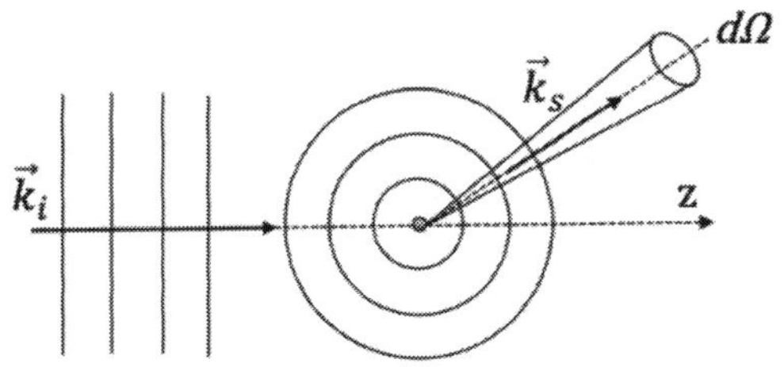

Fig. 2. Representation of the incident planar wave, with the wave number $\vec{k}_i$, scattered at a point scatterer into a spherical wave with the wave number $\vec{k}_s$.

The incident neutron flux is equal to:

$$\Phi = v\|\Psi\|^2 = v$$

The number of neutrons scattered by time unit through *dS*, the surface element (on a detector for instance) defined by the solid angle $d\Omega = \frac{dS}{r^2}$, can be calculated by:

$$vdS\|\Psi'\|^2 = vdS\frac{b^2}{r^2} = vb^2 d\Omega$$

where v is the neutron velocity and $\|\Psi'\|^2$ is the probability of finding neutrons in *dS*.

The number of neutrons *per* solid angle is the differential scattering cross section $\frac{d\sigma}{d\Omega}$ and is calculated by the ratio of the scattered neutrons divided by the incident flux and the solid angle:

$$\frac{d\sigma}{d\Omega} = \frac{vb^2 d\Omega}{\Phi d\Omega} = b^2$$

By integrating on the sphere around the sample (represented by a point scatterer), we find the differential section of a nucleus (the usual unit is barn, equal to 10^{-24} cm^2):

$$\sigma = 4\pi b^2$$

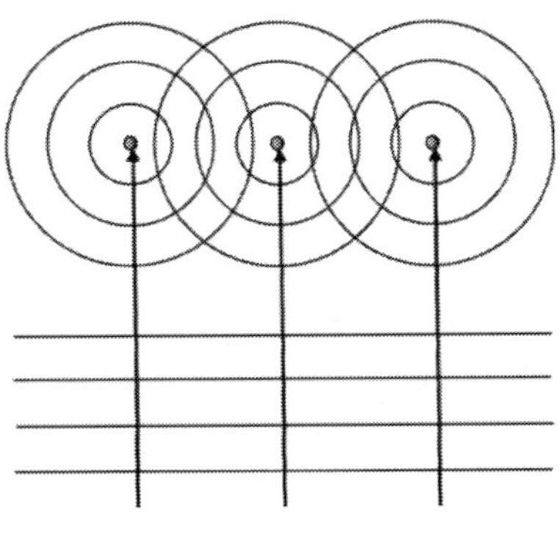

Fig. 3. Principle of a neutron scattering experiment showing interferences between three spherical waves generated by the interaction of the neutron planar wave with three point scatterers representing three nuclei in the sample.

In a neutron scattering experiment, the spherical waves, generated by the sample nuclei from the planar wave of each incident neutron, interfere with each other (Fig. 3). Some interferences between i and j nuclei are, either constructive, or destructive, therefore the spherical symmetry is lost and we observe oscillations. The differential scattering cross section, that makes an average in the ensemble, is written:

$$\frac{d\sigma}{d\Omega} = \frac{1}{N}\langle \textstyle\sum_{i \le j} b_i b_j e^{(i\vec{Q}\cdot\vec{r}_{ij})} \rangle \quad \textbf{Eq. 2}$$

2.4 Coherent/incoherent scattering

The term "coherence" we use for neutrons is not related to a property of the sample coherence (spatial correlation), nor to the neutron beam (which is rigorously incoherent).

In the interaction between a neutron and a nucleus, except the case of isotopes with null nuclear spin (*e.g.* C, O), there are two possibilities: the neutron couples with the nuclear spin of each atom (noted I), either parallel, or anti-parallel. Since the neutron has a spin equal to $\frac{1}{2}$, the spin of the ensemble, for each isotope, is: $I+\frac{1}{2}$ (parallel spins) or $\left|I-\frac{1}{2}\right|$ (anti-parallel spins) with the respective probabilities (weighting factors): $f_+ = \frac{I+1}{2I+1}$ and $f_- = \frac{I}{2I+1}$.

Each interaction process corresponds to different scattering lengths, b_+ and b_-, respectively. We can define the average and the standard deviation, that give, respectively, the "coherent" and "incoherent" scattering lengths:

$$b_{coh} = \bar{b} = f_+ b_+ + f_- b_-$$

$$b_{inc} = \sqrt{\overline{b^2} - (\bar{b})^2}$$

The "coherent" scattering gives information on the spatial and temporal coherences between positions and motions of pairs of atoms nuclei in the sample. The coherent part of the scattered intensity depends therefore on the interaction between pairs of atoms and on collective excitations.

The "incoherent" scattering does not give any information about pair correlations as it depends only on individual nuclei and so gives information about individual (or self) correlations. Self-correlation techniques are able to probe dynamics of *individual* atoms, even in a concentrated sample.

3 The "contrast variation" technique in biology

Neutrons interact with the nuclei of atoms, not with electrons like X-rays do. Because the size of an atomic nucleus is several orders of magnitude smaller than the neutron wavelength, the scattering is isotropic and the scattering length ($b_{nucleus}$), which corresponds to the atomic "scattering power", does not depend on the scattering angle.

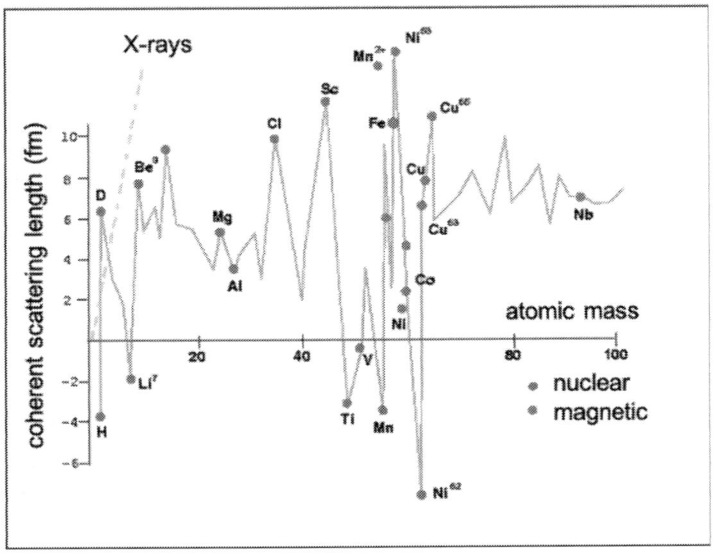

Fig. 4. Coherent scattering lengths of isotopes. Note that there is no relationship with their atomic numbers, in contrast to X-ray scattering lengths.

Contrary to X-ray scattering, the nuclear scattering length of atoms is independent of their atomic numbers (Fig. 4). Therefore, light and heavy elements have comparable scattering lengths. Moreover, different isotopes of a given element scatter neutrons in different ways. Some can exhibit very different scattering lengths and even with opposite signs (the negative sign of b can be interpreted as the change of phase between the scattered and the incident waves).

In particular, this is the case for the coherent scattering length of hydrogen (b_H = -3.74 fm) and deuterium, which is positive and very large (b_D = 6.67 fm), and close to that of other atoms found in biological samples (Fig. 5A). The large difference between the coherent scattering length of these two isotopes is very useful for structural studies of biomolecules, namely in techniques such as neutron diffraction, small-angle neutron scattering (SANS), specular neutron reflectometry (SNR), or neutron spin-echo spectrometry (NSE).

The difference between the scattering lengths, or the scattering length densities (which are calculated by dividing the added scattering lengths of atoms that constitute parts of the scattering entities by the occupied volume), allows the definition of a "contrast".

As mentioned above, in biology, as well as in organic compounds, the remarkably large difference between the coherent scattering length of hydrogen and that of other common atoms (deuterium, carbon, oxygen, nitrogen, phosphorous) allows large contrasts between parts of the sample, if one replaces selectively, hydrogen by deuterium. We have also to take into account the exchangeable H nuclei from the sample with the buffer used, that are those bound to N and O atoms (present in NH, NH2, OH, etc.), and called the "labile" protons, but not bound to C atom (CH, CH2, CH3, etc.). See the book section of Jeffries *et al.* for more details.

Isotopic mixtures of H and D (for example, in hydrogenated and deuterated water) can therefore modulate the scattering length density of parts of a sample or of a solvent, erasing the contrast of such parts, thus enhancing the contrast of other parts. This technique, only possible with neutrons, is called "contrast matching" (Fig. 5). For instance, by mixing light and heavy water (or any other hydrogenous and deuterated solvents or buffers), the scattering length density of the solvent can be varied as desired. In the case of mixtures of compounds having different scattering lengths (hydrogenous and deuterated protein complexes, mixtures of hydrogenous and deuterated lipids, hydrogenous proteins in interaction with deuterated lipid membranes, hydrogenous proteins in the presence of a high concentration of deuterated crowding agents, etc.), it is therefore possible to mask either of the components of the compounds.

This is particularly important in biology where hydrogen is very present (50% of all the atoms in most of biological macromolecules) and practically evenly distributed, which is a great advantage to make structural and dynamical studies using neutron scattering. Often, a neutron scattering measurement of a biological sample consists of measuring correlations between regions of the sample rich in hydrogen compared to other parts of the sample.

The notion of contrast must be applied differently at different scales, because a domain can be taken as homogeneous (thus, described by a single scattering length density) at a large scale, and heterogeneous at smaller scales, in which case the contrast must be defined with higher spatial resolution. Therefore, a large macromolecule can be defined by a single or by different scattering length densities at small or large Q-values, respectively. For example, amino acids constituting a protein can be distinguished or not, depending on the covered Q-range of the experiment.

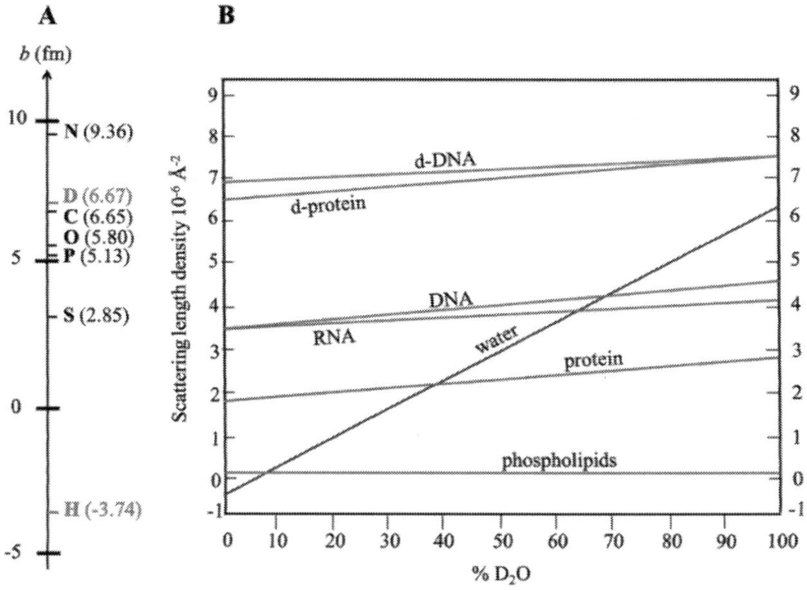

Fig. 5. (A) Coherent scattering lengths, in fm (1 fm = 10^{-13} cm) of the main atoms present in biological molecules. Note that the value of the hydrogen atom (^{1}H) is largely different from those of other atoms, in particular of the deuterium atom (^{2}H = D), which allows high contrasts and is the basis of the "contrast variation" technique. **(B)** Scattering length densities of the natural and of a few fully deuterated biological molecules as a function of the isotopic composition of water. Lines are not horizontal because of labile hydrogen atoms, *i.e.* atoms of the solute that exchange with atoms of the solvent. This graph applies to water and, within good accuracy, to common buffers.

Some isotopes have also very different *incoherent* scattering cross sections. This is also the case of H and D ($\sigma_{inc}(H) \simeq 80$ barn compared to $\sigma_{inc}(D) \simeq 2$ barn). Because of this large incoherent cross section, the motions of hydrogen atoms can be observed directly in quasi-elastic scattering, evaluated by time-of-flight spectroscopy (see 4.2).

4 What type of information does neutron scattering bring to biology?

Several neutron scattering techniques are used in biology to study structure (crystallography [8], SANS (combined with simulations [9], SAXS [10,11], or NMR [12]), and reflectometry [13,14]) and dynamics [15,16], which are presented in the present book. The reader is invited to refer to the following book sections to focus on:

- **Neutron diffraction** ("Opportunities and challenges in neutron crystallography" by N. R. Zaccai and N. Coquelle),
- **Small-angle neutron scattering theory** ("The basics of small-angle neutron scattering" by C. M. Jeffries, Z. Pietras, and D. I. Svergun); **applications** ("The power of SANS, combined with deuteration and contrast variation, for structural studies of functional and dynamic biomacromolecular systems in solution" by E. Mahieu, Z. Ibrahim, M. Moulin, M. Härtlein, B. Franzetti, A. Martel, and F. Gabel), and **simulations** ("Solution structure of macromolecules using small-angle neutron scattering and molecular simulations" by J. Bhatt),
- **Neutron reflectometry theory** ("An introduction to neutron reflectometry" by F. Cousin and G. Fadda) and **applications** ("Applications of neutron reflectometry in biology" by Y. Gerelli),
- **Quasi-elastic and inelastic neutron scattering** ("Probing the dynamics of biological matter by elastic, quasi-elastic, and inelastic neutron scattering" by G. Schirò).

Figure 6 compares the space-time domains covered by different neutron spectrometers compared to other techniques. It is worth noting that methods are complementary and cannot be compared in a simple way. This is because *i*) different probes interact with the sample in different ways and *ii*) measured correlations correspond to individual, pair, or collective quantities.

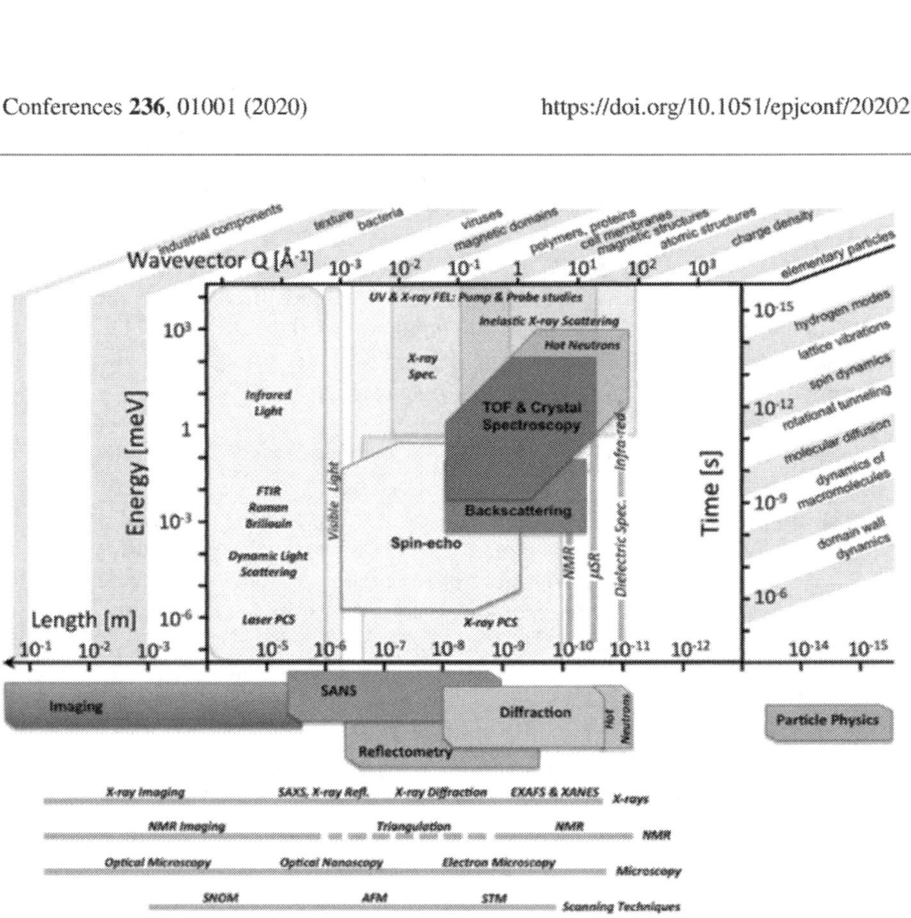

Fig. 6. Complementarity of inelastic neutron scattering, time-of-flight, backscattering, and neutron spin-echo with other spectroscopic techniques. The spatial (r [Å] $\approx 2\pi/Q$ [Å^{-1}]) and temporal (t [ps] $\approx 0.658/\hbar\omega$ [meV]) regions associated with the main physical phenomena potentially observable are shown. One can note that neutron scattering has the capability to simultaneously probe a significant region of this spatial and temporal map. Also, the spatial dependence of the correlation time is directly accessible. From [17] and the European Spallation Source (ESS) web site (https://europeanspallationsource.se/science-using-neutrons).

To give a flavor of the following book sections, we give below a succinct description of both structural and dynamical information neutron scattering can bring to biology, hoping that "appetite will come with eating".

4.1 Structural information

In structural studies (spatial correlations), incoherent scattering generates a constant background that adds to the coherent intensity, the only one containing information about spatial correlations in the sample. This is illustrated by the following equations. The general relation of neutron intensity can be written from Eq. 2 (see 2.3):

$$S(\vec{Q},t) = \sum_{i,j} b_i^{coh} b_j^{coh} \langle e^{[i\vec{Q}\cdot\vec{r}_i(0)]} e^{[-i\vec{Q}\cdot\vec{r}_j(t)]} \rangle + \sum_i (b_i^{inc})^2 \langle e^{[i\vec{Q}\cdot\vec{r}_i(0)]} e^{[-i\vec{Q}\cdot\vec{r}_i(t)]} \rangle$$

When using the static approximation, we obtain:

$$S(\vec{Q}) = \sum_{i,j} b_i^{coh} b_j^{coh} e^{[i\vec{Q}\cdot(\vec{r}_i-\vec{r}_j)]} + \sum_i (b_i^{inc})^2$$

In a totally random sample, the coherent scattered intensity is null. However, in biological samples, like in condensed matter in general, correlations are always present, at least due to steric effects or to compressibility in the case of fluids.

The extension of spatial correlations goes from interatomic distances (*e.g.* atoms of a molecule) to infinity, in a perfect crystal. A particular neutron technique can measure correlations within a relatively small spatial range, reason why different instruments are necessary to evaluate coherences at different scales.

Elastic scattering gives access to the structural properties of matter, *i.e.* the way atoms, molecules or complex structures organize themselves and interact in space. Information is always derived from spatial correlations between sites or spatial domains. The size of such domains goes from 0.1 to 1,000 nm.

High spatial resolution is mostly used to determine crystalline structures by diffraction. Most of the time, **neutron diffraction** complements X-ray diffraction, in particular to determine positions of hydrogen atoms or protons (H^+), which interact too weakly with electromagnetic radiation. The basic principles of neutron diffraction are similar to those of X-rays. In certain directions, the periodicity of the lattice of a crystal induces strong constructive interferences resulting in the observation of Bragg peaks (for more details, especially on diffraction theory, refer to the book section written by Zaccai and Coquelle).

With a much lower spatial resolution, **small-angle neutron scattering** (SANS) probes global sizes and shapes of macromolecules in solution as well as their mutual interactions. Obtained results are very reliable because they correspond to statistical averages over an extremely large number of scattering entities, essentially, all those contained in a sample volume of the order of 100 mm^3. Cryo-microscopy can give equivalent information but statistics refers only to a few molecules and the necessary preliminary preparation of the sample can influence the observations. On the other hand, random inhomogeneity is averaged and is therefore not observed in reciprocal space.

Measuring with high resolution the small Q-domain, SANS of biomolecules in solution can often be decomposed into two independent terms. One of them, dominating the large Q-domain, gives information about size, shape, and surface area of the molecules, while scattering at lower angles gives information about mutual interactions. In this case, scattering is the most precise technique allowing the evaluation of interactions, thus of the effective potentials that define large-scale interactions. Within simple, often valid, assumptions, the total scattered intensity is the product of two terms, *form* and *structure* factors describing molecular shape and interactions, respectively.

Whenever reliable, extrapolation to zero angle of scattered intensity yields information about thermodynamic quantities such as molecular mass or osmotic pressure. Even less accurate SANS data can give information about global quantities such as radius of gyration (R_g) of an aggregate or a complex, or about the area of interfaces. Indeed, at large Q-values, SANS probes correlations at small length scales, *i.e. intra*-particle correlations, while at the smallest Q-values, large correlation lengths characterizing *inter*-particle correlations are the main contribution to the scattered intensity (Fig. 7).

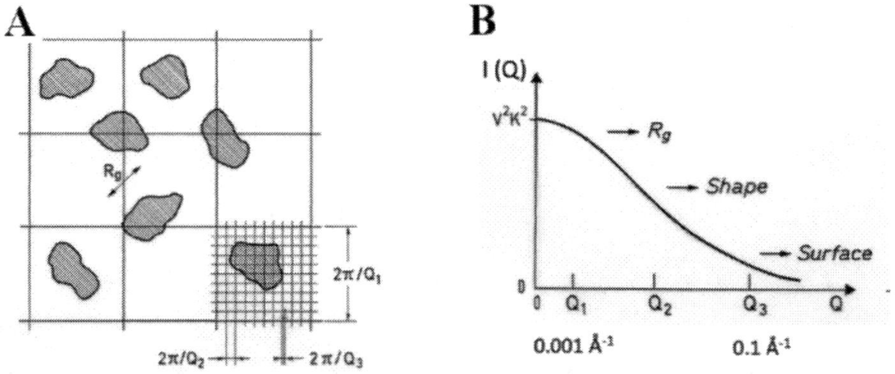

Fig. 7. (**A**) Schematic representation of a two-dimensional sample, showing different degrees of spatial resolution as a function of Q and (**B**) the corresponding regions of the form factor ($I(Q) = P(Q)$) curve. From [1].

In real space, the product of form and structure factors corresponds to the convolution of the self-correlation of the particles with the pair correlation function that describes the spatial arrangements of the particles in the solution.

More details on SANS theory, applications, and simulations can be found in the book sections written by Jeffries *et al.*, Mahieu *et al.*, and Bhatt.

Scattering principles of **specular neutron reflectometry** (SNR) are the same as for SANS but applied to surfaces, instead of volumes. SNR consists on sending a collimated neutron beam onto a surface with a certain incidence angle and measuring the reflectivity of the sample as a function of the vector scattering, noted Q_z, in the direction normal to the surface, and with an amplitude equal to:

$$Q_z = \frac{4\pi}{\lambda} \sin(\theta)$$

where λ is the neutron wavelength and θ the incidence angle (for more details, refer to the book section written by Cousin and Fadda). The reflectivity curve obtained makes it possible to determine the thickness and the scattering length density of each layer of the sample. This is called “specular” reflectivity (like a mirror) as opposed to the “grazing-incidence” (GI-SANS) SANS technique which provides information in the plane area.

4.2 Dynamical information

Neutron scattering uses “thermal neutron” particles that have been thermalized by inelastic exchanges with a thermal bath, either water at room temperature, or liquid hydrogen at 20 K. The resulting kinetic energy of neutrons that interact with matter is similar to that of intermolecular bonds, thermally activated processes, or collective excitations, which are associated to energies smaller than a few tens of meV. Therefore, the energy domain probed by neutrons is very different from that of X-rays or light spectroscopy, for which the characteristic energy of the photons is one million or one thousand times larger, respectively.

Additionally, the interaction with the only atomic nuclei (we exclude magnetic interactions) makes the theory relatively simple because describing only atomic displacements (structural information) and vibrational and collective motions (dynamical information).

As pointed out above, a specific advantage of neutron scattering is the existence of two contributions to the scattered intensity, coherent and incoherent, which can be tuned often by isotopic substitutions that, within good approximation, do not modify physical or chemical properties of the samples. In practice, this remark applies almost exclusively to hydrogen and deuterium atoms, often present in biomolecules and hydration water.

While, in structural studies, the incoherent scattering represents a simple Q-independent background without information about pair correlations, in dynamic studies, it gives information about individual motions of atoms. This property is almost unique in scattering techniques, for which information about individual motions can be obtained but only indirectly from measurements of mutual diffusion coupled with appropriate models (for example, in dynamic light scattering).

However, in practice, measurements of individual motions are possible only for hydrogen atoms. This is because the total coherent scattering intensity is proportional to N^2 while the incoherent part is proportional to N, where N is the number of nuclei in a sample. Only the very large incoherent cross section of hydrogen, typically two orders of magnitude larger than all other nuclei other nuclei found in biological samples ($\sigma_{inc}(H) \simeq 80$ barn compared to $\sigma_{inc}(C, O, N, D) \simeq$ 0-2 barn) can compensate for this difference.

Therefore, in practice, evaluations of individual dynamics are almost exclusively achieved through measurements of motions of hydrogen atoms.

In biology, atomic fluctuations on the ps-ns timescale are of particular interest because they result from and reflect the forces that structure biological macromolecules, including atomic motions and molecular flexibility associated with biological activity.

Studies of dynamics are however not as numerous as structural investigations. Beyond the fact that there are fewer problems related to microscopic dynamics, neutron scattering has several limitations both at the level of the samples and of the covered dynamical range.

Without going into details of each technique (for that, the reader can refer to the book section written by Schirò), we can briefly describe the main classes of techniques and instruments currently used.

4.2.1 Quasi-elastic neutron scattering

Quasi-elastic neutron scattering (QENS), performed in time-of-flight (TOF) spectrometers, measures dynamical processes in the domain 1-100 ps with good accuracy. Through this technique, it is possible to discriminate between rotational and translational motions. It has been used in many studies of dynamics of water in hydrated molecules giving information about self-diffusion, rotational motions, jumps between sites, hydrogen bond lifetime, etc. It probes similar motions in biomolecules, as far as associated relaxation times are not too long as compared to the instrumental resolution, which is of the order of tens of μeV.

Given the characteristic space-time domain better covered by neutron scattering, the majority of QENS studies concerns motions at molecular scale, meaning that collective motions (cf. 4.2.3) often associated to biological functions only indirectly can be taken into account. This is the case of the kinetics of protein folding, which is too long to fit in the experimental window. Instead, dynamical properties of stable refolding intermediates can be measured in order to be compared with those of the native state.

This technique allows a direct measurement of the van Hove function $G_{self}(r,t)$ of the hydrogen atoms present in the sample (the following being related to isotropic systems, we substitute r to $\vec{r}$). $G_{self}(r,t)$ is the probability of finding a hydrogen atom at point r and at time t assuming that the same atom was at the origin at time $t = 0$.

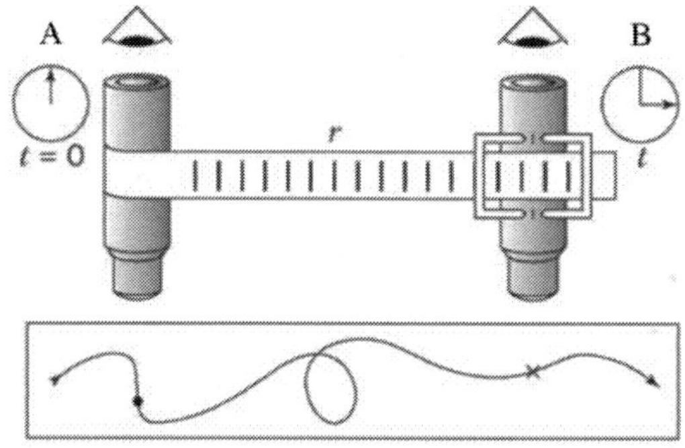

Fig. 8. A "fly correloscope". From [2].

To illustrate the meaning of $G_{self}(r,t)$ with an image, R. Scherm and B. Fåk [2] suggested the image of a "fly correloscope", constituted by two viewfinders A and B connected by a graduated ruler and a chronometer, to illustrate how to measure the probability of having a fly at $(\vec{r}, t)$ if we have seen the same fly at (0,0) (Fig. 8).

Figure 9 depicts schematically the shape of $G_{self}(r,t)$ in the case of a simple diffusion process with spherical symmetry.

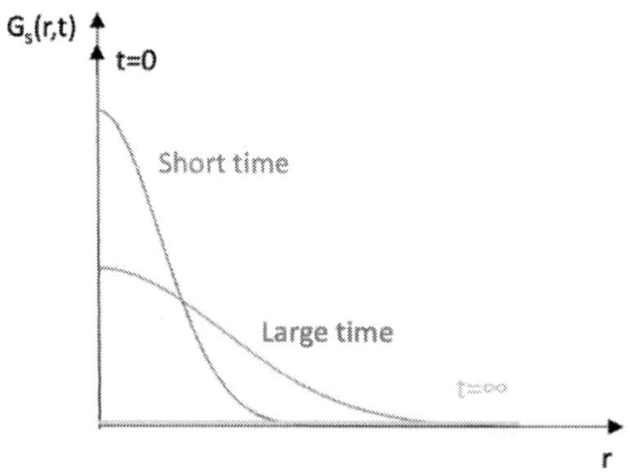

Fig. 9. Shape of $G_{self}(r,t)$ as a function of time.

At $t = 0$, the diffusing particle is at the origin. With increasing time, it can be found in a larger region and, at infinite time, it can be anywhere in the sample. Obviously, all these bell-shaped functions have the same integral, equal to 1.

In practice, the measurement can evaluate diffusive motions of large particles containing hydrogen atoms, as far as internal motions of hydrogen atoms take place at a different time scale. In case of rotational motions, the particle remains confined in a limited region of the space, meaning that $G_{self}(r,\infty) = \text{constant} \neq 0$.

As we have seen above, about the structure factor, neutron scattering analyses correlations in reciprocal space, for which the variable reciprocal of the real space r is the momentum exchange Q. Similarly, concerning dynamics, many instruments measure the energy exchange, $\Delta E = \hbar\omega$, between neutrons and nuclei, ω being the conjugate variable of time t. We illustrate below the relationships between the van Hove function and the spectra measured by neutron scattering:

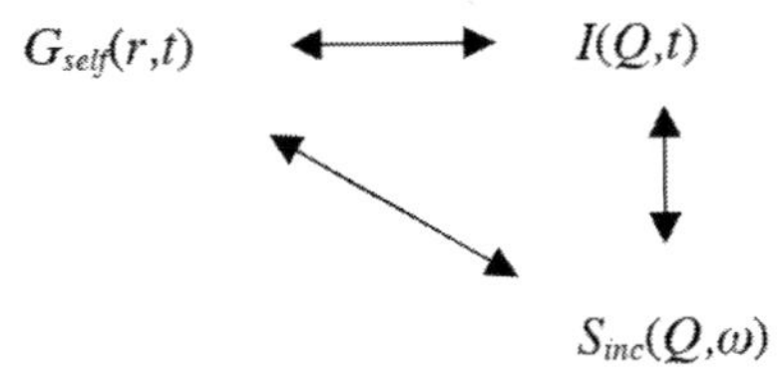

The index *self* refers to individual dynamics, which, most of the time (but not always) is measured by the incoherent part of the scattered intensity, $S_{inc}(Q,\omega)$. Arrows represent space and time Fourier transforms. $I(Q,t)$ and $S(Q,\omega)$ are the intermediate and the dynamical scattering functions, respectively:

$$S(\vec{Q},\omega) = \frac{1}{h}\iint G(\vec{r},t)exp\left[i\left(\vec{Q}.\vec{r} - \omega t\right)\right]d\vec{r}dt$$

Majority of spectrometers measure $S(Q,\omega)$. This is the case of different types of TOF and backscattering instruments. Typical energy resolutions are 50 and 0.1 μeV (FWHM), respectively. Based on a different principle, spin-echo instruments measure with a high energy resolution $I(Q,t)$, *i.e.* relaxation times up to 10 ns (we remind that t [ps] $\approx 0.658/\hbar\omega$ [meV]), typical of the dynamics of large molecules or aggregates.

4.2.2 Inelastic neutron scattering

Inelastic neutron scattering (INS) probes atomic and molecular vibrational modes, namely the vibrational density of states related to thermodynamic properties and phase transitions. They are mostly in the intermolecular energy domain below 100 meV.

They are measured in TOF spectrometers, often with low energy resolution in order to reach high energy exchange at moderate Q values.

Actually, as it can be seen in Figure 10, there is a strong interdependence between Q and ω, because of the small energy of neutrons. The neutron dispersion curve, that follows Eq. 1 (see 2.2), relates Q and ω for given incident neutron wavelength $\lambda = 2\pi/k_I$ and scattering angle θ.

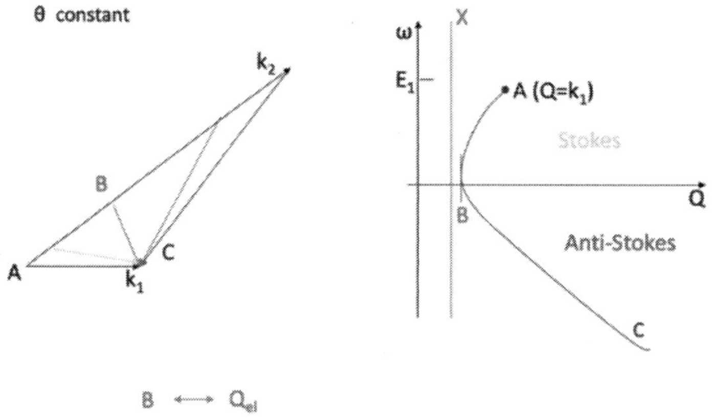

Fig. 10. The neutron dispersion curve relying the momentum Q and the energy ω, for a given wavelength λ and a scattering angle θ.

On the same scale, the equivalent curve for X-rays or light scattering is reduced to a vertical line because the incident energy is much higher than the energy exchanges probed by neutrons.

One consequence of the neutron dispersion is the small extension of the Stokes side (neutron energy loss) of the spectra. Therefore, analysis is done on the anti-Stokes side (neutron energy gain) despite the lower intensity, particularly at low temperature.

Still, INS is a very useful probe that, in biology, often complements light scattering in the difficult domain of far infra-red absorption (typically up to a few THz, with 1 THz = 32 cm^{-1}) where intermolecular contributions dominate.

4.2.3 Collective motions

More rarely in biology, INS can probe collective excitations, namely propagative excitations and acoustic modes (sound waves). Equivalent of Brillouin light scattering in a domain of frequency three orders of magnitude larger (THz instead of GHz) the experiments evaluate the energy dependence of the coherent part of the scattering function, $S_{coh}(Q,\omega)$ in the hydrodynamic limit, *i.e.* at small Q-values. Technically, experiments are difficult because only a narrow domain of Q and wavelength is available and the scattering intensity is always weak. This is because typical values of sound velocity imply using short wavelength neutrons, hardly compatible with small-angle scattering.

Today, high resolution X-ray inelastic scattering is the main method in this domain of research.

4.2.4 Elastic neutron scattering

Strictly speaking, elastic scattering is solely due to immobile atoms or to local motions taking place at scales relatively small. Examples of the latter are intramolecular rotations of methyl groups or diffusion under high confinement.

The eventual Q-dependence of the intensity of the elastic line gives information about the geometry of the confined volume where motions, namely rotations, take place. An important quantity often evaluated from the elastic component is the elastic incoherent structure factor (EISF) that gives information about the size of a domain where confined motions take place. Therefore, although indirectly, it gives structural information about the sample.

However, in case of slow diffusion, more generally in case of long relaxation times, the instrumental resolution is not good enough to discriminate slow motions from real immobility. In practice, what is measured like "elastic" scattering, does include motions associated to long relaxation times. In elastic scans, one measures the intensity scattered inside a dynamic window generally centered at $\omega = 0$. In general, its width is the instrumental resolution, eventually larger. Measurements are performed in function of an external parameter, normally the temperature.

In the case of a sample in which all atoms are immobile, one will observe a regular decrease of the "elastic" intensity due to the Debye-Waller factor, which describes atomic delocalizations. With increasing temperature, part of the intensity is scattered inelastically, reducing the intensity scattered elastically. Actually, beyond this trivial effect, one looks for kinks or discontinuities of the elastically scattered intensity plotted against temperature. They correspond to dynamic changes, such as onset of dynamic processes, glass or phase transitions.

It is worth noting that the most frequent dynamics of biomolecules is associated to relaxation times larger than 1 ns, which are too large to be studied in TOF spectrometers. Therefore, spin echo instruments and elastic scans, often performed in backscattering instruments, are currently used.

Many other techniques are possible with neutrons in biology that are not described in the present book (*e.g*. imaging). We did not give also technical details on neutron sources and instruments (for that purpose, refer to [2,4]).

Nevertheless, we hope that we convinced the reader that neutron techniques are useful probes in many cases in biology and that this introduction encourages to go further, which is also the propose of the following chapters of this book, that we invite to discover.

References

1. J. Teixeira, *Introduction to small angle neutron scattering applied to colloidal science*, 1992, Struct. Dyn. Supramol. Aggregates Strongly Interact. Colloids, NATO ASI S, p. 635.
2. R. Scherm, B. Fåk, 1993, Neutron Synchrotron Condens. Matter Stud., Springer-V, Les Ulis, p. 113.
3. J. Schweizer, 2005, Neutrons Biol., Journal de Physique IV.
4. T. Ederth, *Neutrons for scattering: What they are, where to get them, and how to deal with them*, 2018, JDN 23 - French-Swedish Winterschool Neutron Scatt. Appl. to Soft Matter, EPJ Web of Conferences.
5. C.M. Jeffries et al., Nat. Protoc. **11**, 2122 (2016)
6. D. Franke et al., J. Appl. Crystallogr. **50**, 1212 (2017).
7. J. Trewhella et al., Acta Crystallogr. Sect. D Struct. Biol. **73**, 710 (2017)
8. M.T. Banco et al., Proc. Natl. Acad. Sci. U. S. A. **113**, 13756 (2016)
9. R. Dos Santos Morais et al., Biophys. J. **115**, 1231 (2018)
10. F. Spinozzi et al., Langmuir **33**, 2248 (2017)
11. M. Chevreuil et al., Nat. Commun. **9**, 3071 (2018)
12. A. Lapinaite et al., Nature. **502**, 519 (2013)
13. A. Martel et al., J. Am. Chem. Soc. **139**, 137 (2017)
14. S. Combet et al., Biochim. Biophys. Acta - Biomembr. **1861**, 514 (2019)
15. G. Schirò and M. Weik, J. Phys. Condens. Matter. **31**, 463002 (2019)
16. S. Combet and J.-M. Zanotti , Phys. Chem. Chem. Phys. **14**, 4927 (2012)

17. Q. Berrod et al., 2018, *Inelastic and quasi-elastic neutron scattering. Application to soft matter.* French-Swedish Winterschool Neutron Scatt. Appl. to Soft Matter, EPJ Web of Conferences.

EPJ Web of Conferences **236**, 02001 (2020) https://doi.org/10.1051/epjconf/202023602001
JDN 24

Opportunities and challenges in neutron crystallography

Nathan Richard Zaccai[1,*], and *Nicolas* Coquelle[2,†]

[1]CIMR, University of Cambridge, Cambridge CB2 0XY, United Kingdom
[2]Institut Laue Langevin, 38042 Grenoble Cedex 9, France

Abstract. Neutron and X-ray crystallography are complementary to each other. While X-ray scattering is directly proportional to the number of electrons of an atom, neutrons interact with the atomic nuclei themselves. Neutron crystallography therefore provides an excellent alternative in determining the positions of hydrogens in a biological molecule. In particular, since highly polarized hydrogen atoms (H+) do not have electrons, they cannot be observed by X-rays. Neutron crystallography has its own limitations, mainly due to inherent low flux of neutrons sources, and as a consequence, the need for much larger crystals and for different data collection and analysis strategies. These technical challenges can however be overcome to yield crucial structural insights about protonation states in enzyme catalysis, ligand recognition, as well as the presence of unusual hydrogen bonds in proteins.

1 Introduction

Although X-ray crystallography has become the workhorse of structural biology, Neutron crystallography has several advantages to offer in the structural analysis of biological molecules.

X-rays and neutrons interact differently with matter in general, and with biological macromolecules in particular. These two crystallography approaches are therefore complementary to each other [1]. While X-ray scattering is directly proportional to the number of electrons of an atom, neutrons interact with the atomic nuclei themselves. In this perspective, hydrogen atoms, which represent ~50% of the atomistic composition of proteins and DNA, are hardly visible using X-ray crystallography, while they can be observed in nuclear density maps derived from neutron diffraction data even at moderate resolution (2.5Å and higher). In fact, less than 5% of deposited models in the Protein Data Bank (PDB) were obtained from crystals diffracting X-rays to (sub-)atomic resolution, and even in the structures with resolution better than 0.5 Å, a third of the expected hydrogens could still not be experimentally identified [2]. The positions of the more mobile or labile hydrogen atoms, which are typically the most biologically relevant, had to be inferred. Neutron crystallography therefore provides an excellent alternative to determine the position of

* Corresponding author: nrz20@cam.ac.uk
† Corresponding author: coquelleni@ill.fr

hydrogen atoms in a molecular structure. It is even able to identify highly polarized H atoms or protons (H^+), which cannot be observed by X-rays as they do not have any electrons.

This neutron specificity arises from the coherent scattering length of the two stable isotopes of hydrogen (^1H and ^2H (also delineated deuterium, D)) being of similar magnitude to that of other atoms which compose a biological macromolecule (Figure 1). Deuterium is mentioned here as it has similar chemical properties to the Hydrogen atom, as well as physical properties more favourable for neutron diffraction experiments. H/D isotopic replacement within the crystal (where Hydrogens are exchanged for Deuterium atoms) is mandatory for experimental success.

Visualization of hydrogen and deuterium atoms can give important insights regarding the chemistry of the biological system. For example, knowledge of the protonation states of residues within the catalytic site of an enzyme is crucial to understand the catalytic mechanism involved [3–8]. Positioning hydrogen atoms unravel the exact hydrogen-bond network that usually govern ligand or drug binding to biomacromolecules [9–14]. The hydration pattern also plays a significant role in the thermodynamic process of ligand binding [15]. Water molecule orientations can be directly inferred from neutron diffraction data. Indeed, waters are often present in protein active sites and form an integral component of the catalytic reaction [16,17].

A significant advantage of neutrons is that, since their scattering is non-destructive, data collection can be performed at room temperature. Obtaining a structure free of damage is of great interest for biological systems that are highly sensitive to X-ray radiation damage [18]. For example, redox centres of metallo-proteins can be damaged by X-rays even at temperatures as low 100K [19,20]. Neutron diffraction has also been used to identify the chemical nature of catalytic intermediates of heme-peroxidases [7,8], while dozens of X-ray datasets could not converge onto a unique structural model due to radiation damage of the active site.

While neutron crystallography is a unique technique, it bears its own limitations, due to a small number of available instruments (when compared to X-rays), the inherent low flux of neutrons sources (order of magnitude smaller than an X-ray rotating anode), and as a consequence, the need for much larger crystals. As of December 2019, only 161 neutron structures were available in the PDB. However, progress made on sample preparation (perdeuteration and large crystal growth apparatus) and on instrumentation have moved the field forward. With more than half of all neutron models deposited within the last 5 years, most, if not all, now provide important answer(s) to specific biochemical question(s), that will ultimately lead to the need to rewrite biochemistry textbooks [6,7].

In this chapter, we will describe the challenges that neutron crystallography needs to face, the instrumentation and different crystallographic method used at various sources (reactor versus spallation), the data reduction step and model refinement, and finish with detailed examples of biological questions which can be addressed using neutron crystallography.

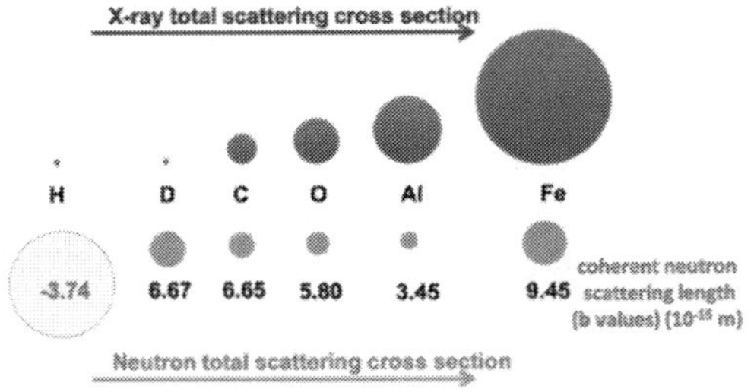

Fig. 1. Scattering power difference between X-rays (blue circles) and neutrons (yellow and green circles). Please note that the surface corresponds to the total, *i.e* coherent (which contributes to signal) and incoherent (which contributes to the background) scattering lengths. While the coherent scattering lengths are in the same range for all elements found in the composition of biological molecules, the much larger surface for the hydrogen element comes from its large incoherent cross section.

2 Neutron crystallography challenges

The basic principles of neutron diffraction are similar to that of X-rays. The repeating motif of the crystal, called unit cell, yields in certain directions strong constructive interference, resulting in a Bragg diffraction spot. In other directions, interferences are destructive and no signal is measured. Each Bragg diffraction spot can be annotated to a particular (hkl) index, (Miller index) indicative of the diffracting plane. The measured intensity of a given hkl plane is determined by the corresponding structure factor *F(hkl)* and is expressed as:

$$\boldsymbol{F(hkl)} = \sum_j b_{c,j}\, e^{2\pi i(hx_j+ky_j+lz_j)} e^{-B_j\frac{sin^2\theta}{\lambda^2}} \tag{1}$$

The summation is performed over all the atoms within the unit cell. Here, $b_{c,j}$ and B_j represent the coherent scattering length and the temperature factor of the atom *j* [21]. In the case of X-rays, the formula is equivalent, with the exception of the form factor f_j in place of the coherent scattering length. The goal of a diffraction experiment is to position the atoms in the unit cell. This is done through the computation of the nuclear density, which is the Fourier transform of the structure factor *F(hkl)*:

$$\rho(xyx) = \frac{1}{V}\sum\sum\sum_{h,k,l} \boldsymbol{F(hkl)}\, e^{-2\pi i(hx+ky+lz)} \tag{2}$$

where $\rho(xyz)$ is the nuclear density at position (x,y,z) in the crystal unit cell and V its volume. The structure factor is a complex number with its real part, the structure factor amplitude measures the magnitude of the diffracted wave, and its imaginary part, the phase of the diffracted wave:

$$\boldsymbol{F(hkl)} = |\boldsymbol{F(hkl)}|\, e^{i\,\alpha_{hkl}} \tag{3}$$

The structure factor amplitude is proportional to the measured intensity in a neutron diffraction experiment, while the phase α remains unknown from such experiment. The delineated “phase problem” will be mentioned later on. For now, we focus on the measured intensity, and its related experimental parameters. Indeed, the Bragg diffraction intensity of a spot with indices *hkl* is, in a neutron diffraction experiment:

$$I(hkl) = \phi(\lambda)\,\frac{V_c}{V_{cell}^2}\,F(hkl)^2\,\frac{\lambda^4}{2sin^2\theta} \tag{4}$$

where $\phi(\lambda)$ is the differential neutron flux at the sample, *F(hkl)* is the factor structure amplitude of the reflection, with Bragg angle 2θ, and V_c and V_{cell} are the crystal and unit cell volumes, respectively [22]. Importantly, the intensity of each spot does not directly contain the phase associated with the spot.

While there is a real gain in intensity to increase the wavelength λ from (4), it should also be kept in mind that the highest resolution attainable with an incident beam of wavelength λ is λ/2, and the typical covalent bond lengths shorter than 2 Angstrom. Longer wavelengths will also generate larger blind zone [23], and therefore would require multi-axis goniometry [24]. Because the air scattering and absorption is not as problematic with neutrons as it is with X-rays, longer wavelengths can be used for neutron diffractions experiments (from 2Å, up to 10Å). However, high resolution Bragg spots are backscattered for wavelengths above 2 Å, and a specific detector geometry needs to be envisaged, with a maximal detector coverage to record as many Bragg spot intensities as possible within a single exposure.

By rewriting equation (1), the diffraction intensity from a crystal can be seen as being proportional to:

$$I_0 \times A \times \frac{V_c}{V_{cell}^2} \tag{5}$$

where I_0 is the incident intensity, A is the detector area subtended by the crystal and as previously V_c is the volume of the crystal within the beam, and V_{cell} is the volume of the unit cell. The diffracted beam intensity is therefore inversely dependent on the square of the volume of the unit cell (V_{cell}), meaning that the bigger the unit cell, the larger the crystal needs to be. The other values can however be optimized experimentally: by increasing the neutron flux, by increasing the detector area and by using larger crystals. Although X-ray crystallography has become the workhorse of structural biology, Neutron crystallography has several advantages to offer in the structural analysis of biological molecules.

3 Neutron instrumentation

The recent development and increased performance in neutron protein crystallography was made possible thanks to improvement in instrumentation, drastically reducing the overall data collection time and crystal volume needed for successful neutron diffraction experiments (see table 1). At reactor neutron sources, both monochromatic and quasi-Laue techniques have been developed, with a preferential use of image plates to detect neutrons. At spallation sources, the Laue time of flight method takes advantage of the pulsed nature of the neutron beam.

Table 1. Current list of available and upcoming neutron protein diffractometers. BIX-3 and BIX-4 are currently offline.

Source	Instrument	Location	Reference and Website
Reactor	LADI-III	ILL, France	[25] https://www.ill.eu/users/instruments/instruments-list/ladi-iii/description/instrument-layout/
	D19	ILL, France	https://www.ill.eu/users/instruments/instruments-list/d19/description/instrument-layout/
	DALI	ILL, France	
	IMAGINE	HFIR, USA	[26] https://neutrons.ornl.gov › imagine
	BioDiff	FRM-II, Germany	[27] https://www.mlz-garching.de/biodiff
	BIX-3	JRR-3, Japan	[28] https://jrr3uo.jaea.go.jp/jrr3uoe/instruments/index.htm
	BIX-4	JRR-3, Japan	
Spallation	MaNDi	SNS, USA	[29] https://neutrons.ornl.gov › mandi
	iBIX	J-PARC, Japan	[30] http://j-parc.jp/researcher/MatLife/en/instrumentation/ns_spec.html#bl03
	NMX	ESS, Sweden	https://europeanspallationsource.se/instruments/nmx

3.1 Weak neutron flux

Nuclear reactors provide a continuous flux of neutrons, which are produced by nuclear fission of ^{235}U. Each fission event produces 3 neutrons on average and releases about 180 MeV, mostly in the form of kinetic energy. It is also possible to produce neutrons *via* the process of spallation. Accelerated particles (generally protons) are bombarded onto a heavy-metal target and, if the incident particle energy is above a given threshold (typically 5-15MeV), neutrons are extracted from the heavy metal nuclei, (which is transmuted into another chemical element). Both these methods produce neutrons of high kinetic energy, which need to be slowed down to become useful for scientists (*i.e* useful wavelengths). Neutrons are therefore moderated by repeated collisions with a hydrogen rich liquid (H_20, H_2 or CH_4). The resultant thermal flux produced by the research facilities sources (steady states or pulsed sources) is about 10^{15} n.cm^{-2}.s^{-1}, which are several orders of magnitude weaker than available laboratory X-ray sources (Figure 2). Combined with the fact that neutrons diffract very weakly, measurement times range therefore from day(s) to weeks at neutron sources compared to less than a few minutes, or seconds at modern synchrotrons.

Though the neutron flux is limited, other technical aspects can be optimized for neutron diffraction experiments.

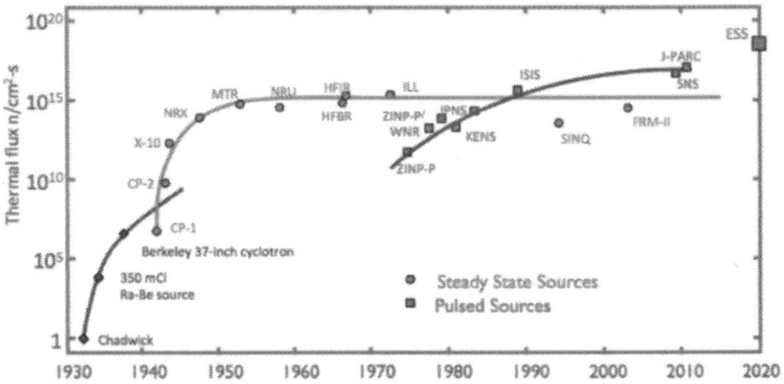

Fig. 2. Representation of the effective thermal flux of various different neutron sources since the discovery of the neutron in 1932 by James Chadwick. Early sources are shown as red diamonds, reactor sources as green circles and spallation sources as red squares. An estimated flux for ESS is indicated by the larger red square. Beware that these are not the fluxes available for the instruments. Reproduced from [31].

3.2 Diffractometers at reactor sources

The continuous neutron flux of the reactor sources is an advantage for protein crystallography as the flux is directly proportional to the measured diffracted intensity (see equation (4)).

At these sources, only the desired neutron wavelengths are selected from the neutron spectrum produced by the reactor (called white beam) by narrow (monochromatic diffractometers) or wide (quasi-Laue diffractometers) bandpass wavelength filters. Monochromators are single crystals, such as pyrolytic graphite, which diffracts a single neutron wavelength from the incident white beam. Wide-bandpass filters are designed to select a wavelength band (~30% for LADI at ILL for example using Ni-Ti supermirrors) which is used for quasi-Laue method. In such case, a large number of the reciprocal lattice points are in diffraction condition within a single crystal exposure (see Figure 3). These methods also provide large gains in flux relative to monochromatic methods, and reduce background scattering and reflection overlap compared to a Laue method (which uses the full white beam). In such experimental setup, the crystal remains stationary during the exposure, and is rotated several degrees between two expositions (5-7 degrees for a 30% wavelength filter). In monochromatic neutron techniques, the wavelength bandpass is only few percent and either the crystal is rotated during the exposure (as with X-rays) or it remains stationary. The angular rotation between two expositions is much smaller (1 degree or less). The vast majority of neutron protein diffractometers at reactor sources (Table 1) uses image plates as neutron detection systems. These are X-ray image plates, which have been doped with a neutron converter (usually Gadolinium) mixed with photo-stimulated luminescence materials [32,33].

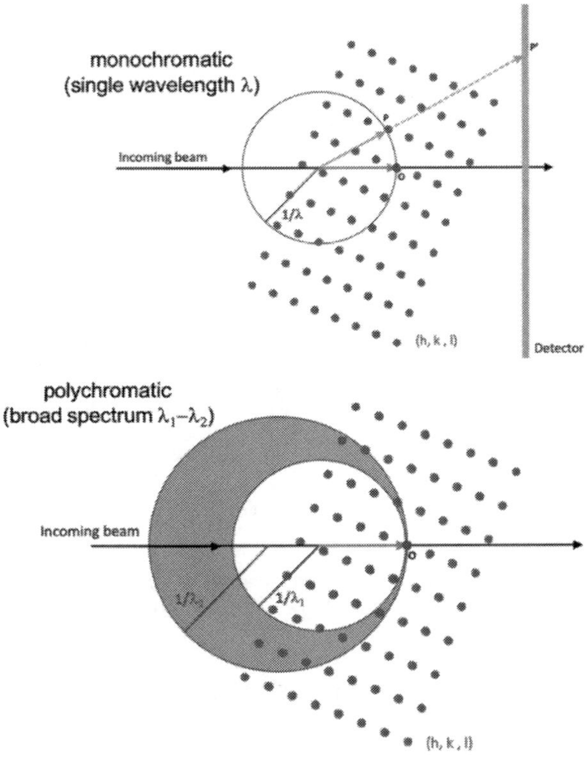

Fig. 3. Ewald sphere construction for monochromatic and quasi-laue diffraction. The Ewald construction is a geometric representation of the diffraction conditions. For neutrons of wavelength λ, a sphere of radius 1/λ is centered on the crystal. Each point or the diffraction lattice has 3 dimensional coördinates denoted by indices (*hkl*). The origin of the lattice (0,0,0) is placed at the intersection of the Ewald sphere with the incident beam. In order for diffraction conditions to be satisfied, the individual reciprocal lattice point has to be on the Ewald sphere. Therefore, by rotating the crystal, and consequently the reciprocal lattice, different diffraction conditions appear. In quasi-Laue diffraction, the crystal is exposed to different wavelength neutrons. The Ewald sphere is consequently broadened out, and more diffraction conditions are satisfied at one crystal orientation.

Importantly, at the usual wavelength used (2.5Å and above), diffracted neutrons are back-scattered. Consequently, contrary to the planar detectors used in X-ray crystallography, a cylindrical shape has been adopted for neutron detectors. Such geometry therefore provides high coverage of reciprocal space (> 2π steradian), and therefore a large number of Braggs reflections can be recorded simultaneously. Although quasi-Laue techniques provide many advantages, increasing the crystal unit cell dimensions, reflections start to overlap, as reciprocal lattice points move closer to each other, thereby limiting the unit cell dimensions accessible to these diffractometers.

Furthermore, DALI, a novel quasi-Laue instrument, with a much narrower wavelength bandwidth (~10% at 3.8Å) will be installed in early 2020 at the ILL. The use of velocity selector to select the usable wavelength from the white beam should also increase the flux 2.5 fold at the sample position.

3.3 Diffractometers at spallation sources

At spallation neutron sources, large position-sensitive detectors (PSDs) allow wavelength-resolved Laue patterns to be collected using all the available neutrons. The diffractometers are specifically designed to use the full white beam. Thanks to the pulsed nature of the neutron beam, not only is the position of a diffracted neutron recorded on the detector, but also the time at which the neutron hits the detector. The addition of this extra dimensions allows the full range of available wavelengths to be used [34]. The Time of Flight (TOF) Laue method therefore has all of the advantages of quasi-Laue methods at reactor sources, but yet does not suffer from overlapped reflections and from background scattering over the wavelength range (the background as well is spread over time bins).

Two TOF Laue diffractometers dedicated to macromolecular crystallography are currently available: MaNDi [35] at the Spallation Neutron Source (SNS) at Oak-Ridge National Laboratory (ORNL) and iBIX [30] at the Japan Proton Accelerator Research Complex (J-PARC). Two new TOF Laue diffractometers should be built within the next few years, NMX at the European spallation source (ESS) and Ewald at the second target of SNS at ORNL [36]. Specific detectors have been built to allow to detect the position and the arrival time of a neutron onto the detector. These position-sensitive detectors (PSDs) have a rather small area of detection and an array of them is positioned around the sample, in a spherical fashion.

Although the TOF Laue technique seems to only present advantages, the pulsed structure of the neutron flux decreases the integrated available flux reaching the sample. With the construction of the next generation spallation source ESS and its integrated neutron flux similar to the neutron flux available at the ILL, this limitation might not hold.

4 Sample preparation for neutron crystallography

4.1 Large crystal growth

A single crystal of the biological sample is essential for crystallography. In order for crystallization to happen, the biological sample must be pure (>95%) and homogeneous. The final step of the sample purification is usually performed *via* size exclusion chromatography in order to remove aggregates (since these would prevent crystallization) and to isolate a single monodisperse species. The biological sample also has to be sufficiently concentrated (typically >10mg/ml for a protein).

Crystal growth can be understood as a controlled precipitation, as molecules associate with each other not randomly as in an aggregate, but in a repeating 3-dimensional motif. Crystallization is driven by variations in precipitant, ionic conditions and pH. In order to identify the optimum crystallization conditions, multi-factorial crystallization screens with over 1000 conditions per biological molecule are typically trialed (see [37]). Salts and different lengths PEGs are good crystallization agents. To understand the crystallization process, a phase diagram is often drawn (Figure 4). On the x-axis is the crystallization agent concentration, while the protein concentration is on the y-axis. The phase diagram is divided into three regions. In the undersaturated region (at low sample concentration), the protein remains soluble. Vapor diffusion techniques (by sitting or hanging drop) will increase the protein and precipitant concentrations to reach the supersaturated region (at higher sample concentration), in which nucleation occurs and crystals appear. In order for larger crystals to grow, upon the onset of nucleation, the soluble protein concentration should drop into the metastable phase, where nucleation cannot occur, but where the crystals can continue to grow. Since nucleation does not occur in the metastable region, it is also possible to use already grown crystals. These are transferred to a solution in which the protein is already in the metastable phase and will only contribute to crystal growth (termed macro-seeding). Other approaches exist, such as batch-method, where the system (protein and precipitant) are mixed to directly reach the supersaturation region [38].

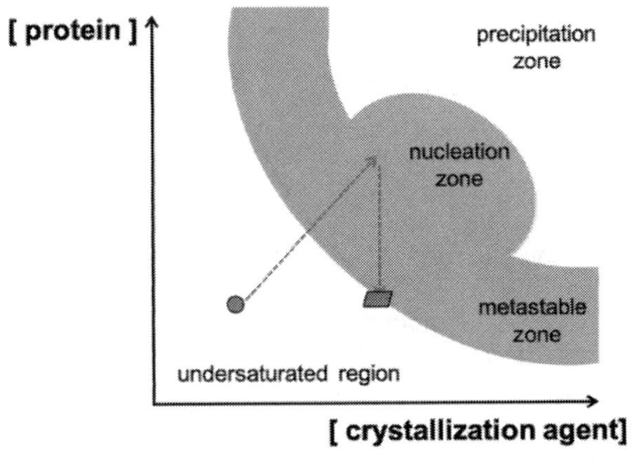

Fig. 4. Crystallization phase diagram. A phase diagram comprises different zones, which either need to be reached (nucleation zone) or avoided (precipitation zone). A typical crystallization starts in the undersaturated region (red dot). As water evaporates from the crystallization drop, both crystallization agent and protein concentrations increase, until reaching the nucleation zone. At this point, protein molecules arrange themselves in an ordered manner, that will lead to a crystal. From there, the protein concentration starts to decrease as the crystal grows. The system reaches equilibrium when the protein concentration is at the limit between the metastable zone and the undersaturated region. The crystal has then reached its maximal size.

Niimura and Bau suggest a detailed crystallization phase diagram should be established [39], but this approach remains time and sample consuming. A possible approach is to use instruments designed to finely control the different parameters of the crystallization experiment, allowing to diffuse to the protein more or less precipitant, as well as provide control of the temperature [40].

As soon as the crystallization conditions have been found and optimized for X-ray determination (which requires only crystals of few tens of microns in size), a second step of optimization is required to grow crystals suitable for neutron diffraction. It is important to emphasize that crystal size is a significant hurdle in neutron crystallography. Put bluntly, bigger crystals are necessary. Several tricks are used to optimize crystal growth. In particular, if crystallization happens too fast, the drop may fill up with a shower of small crystals. In order to slow down crystallization, larger drops can be tried (as they will take longer to equilibrate). Another option is to increase the temperature, until the crystals dissolve, and then reduce the temperature just enough that only a few crystal nucleation events occur, and therefore few (tentatively larger) crystals grow. Small molecule additives can also be incorporated in the previously identified crystallization buffers used to grow the smaller crystals analyzed by X-ray diffraction. Also, we must note that for large crystals, it is not impossible to have the visual impression of a single crystal at the macroscopic scale, which turns out to be slightly differently oriented crystals at the nanoscopic scale. In return, each crystal will concurrently contribute diffraction data, yielding a diffraction pattern of poor quality.

The crystal size needed for a neutron diffraction experiment not only depends on the biological system itself, and its unit-cell parameters, but also depends on the neutron diffractometer performances (linked to the available neutron flux). Hydrogenated crystals should be in the mm^3 range, while one order of magnitude in volume can be gained with a fully perdeuterated crystal. Furthermore, the crystallization buffer surrounding the crystal at the time of data collection needs to be deuterated. Two possibilities, i) the crystal is directly grown in a deuterated buffer or ii) the crystallization solution is hydrogenated and the crystal is subsequently transferred into a deuterated buffer (this can be directly done in the quartz capillaries).

For a given project to be considered for the limited available neutron beam time, crystal structures are systematically solved with X-rays first. This not only provides a starting model on which to append hydrogens later on, but also is the first proof of the diffracting quality of a given protein crystal. For example, some large crystals do not diffract at all; others have unit cells sizes that create challenges for current neutron diffractometer designs (*vide infra*). Also, users will have to show the capacity of their crystal to diffract in a neutron beam. Indeed, X-ray beams are getting smaller and smaller, now being typically in the 50-100μm range (if not smaller), and therefore only impinge a relatively small volume of a large crystal grown for a neutron diffraction experiment. So, X-rays only reveal the diffracting quality of a crystal within this length scale, not through the whole crystal. Neutron beams are much wider (in the mm range), the goal here being to illuminate the complete volume of the protein crystal. In this situation, the crystals need to be ordered on the nanoscopic scale over its entire volume, and bad surprises can arise when a good X-ray diffracting crystal is illuminated with a neutron beam. This is an important point to keep in mind when requesting beamtime for neutron diffraction experiment.

4.2 H/D exchange and perdeuteration

For a successful neutron diffraction experiment, the isotopic replacement of H into D is not just highly encouraged, but is mandatory. It provides two major advantages. First, as the incoherent cross section of ^{1}H is extremely large (80.27 barns) and hydrogen atoms represent ~50% of all the atoms in a protein, their incoherent scattering signal highly contributes to background, and consequently reduce the quality of the recorded diffracted intensities. Deuterium has a 40-fold reduced incoherent cross section (2.05 barns, Figure 1) when compared to hydrogen, and therefore H/D exchange significantly improves signal-to-noise ratio. Second, deuterium has a positive coherent cross section, twice as large (in amplitude) as the hydrogen one, which is negative. Therefore, deuterium is more easily observed in nuclear density maps of moderate resolution (~2.5Å). The negative neutron scattering length of hydrogen, which is opposite in sign to any neighbouring carbons (for example in an aliphatic group), leads to a partial cancellation of the hydrogen and carbon densities, resulting in difficulties to interpret nuclear density maps at these moderate resolutions. Near-atomic resolution (~1.5 Å) is needed for the hydrogens to be visualized in their negative nuclear densities, separated from the positive carbon densities. Finally, the significant solvent content of protein and DNA crystals (50% on average) makes substitution of H_2O into D_2O essential.

Two solutions are available regarding this H/D exchange; either only hydrogens at exchangeable sites can be swapped with deuterium, or the sample itself is fully deuterated (at the time of production by the expression system). With the first option, only a quarter of all hydrogen atoms composing the biomolecule are titrable (plus of course all solvent molecules). The other hydrogens, frequently belonging to methyl and methylene groups, are not. The titrable hydrogens can either be exchanged through vapour H/D exchange when the crystal is mounted in a capillary prior to data collection, or when crystals are soaked in heavy water (D_2O) buffers, (although there is a risk the crystals are damaged during the transfer). Crystals can also be grown directly from deuterated buffers.

The alternative is to incorporate deuterium atoms at the time of protein synthesis. As deuterium has a relatively small incoherent cross-section, crystallographic data of fully deuterated protein exhibit up to 10-fold increase in signal-to-noise ratio in the Bragg peaks in comparison to a H/D exchanged crystal. As a consequence, a crystal ten times smaller in volume can be used.

Large scale neutron facilities now have their own deuteration labs (D-Lab at ILL, Bio-deuteration Laboratory at ONRL, Isis Deuteration Facility, DEMAX at ESS and MLF at J-Parc) which can be helpful for users to produce the deuterated version of their macromolecule of interest. Additionally, it is worth noting that these facilities produce deuterated lipids, sugars, and small molecules.

4.3 Cryo-crystallography

While the non-damaging nature of neutrons permit data collection at room temperature, neutron protein diffractometers have been equipped with the possibility to collect data from frozen crystals. Data collection at cryogenic temperatures offers several advantages, including the decrease in atomic displacement parameters (ADPs, *vide infra*) and improvements in nuclear density definition [41]. The cylindrical image plate detector efficiency also increases at lower temperature.

Importantly, temperature-sensitive crystals can be cryo-cooled prior to neutron data collection. While some tricks can be envisaged to trap enzymatic intermediates at room temperature, usually *via* the design of catalytically inactive variants of the biological molecule of interest, or the use of substrate analogs which block the catalytic reaction at a given step, the most interesting, but most challenging is to trap enzymatic intermediates to identify its true chemical nature. Such strategy was successfully performed on heme-peroxidases, which allowed the determination of so-called compound I [7] and compound II [8] within this family enzyme, unraveling chemical details to define a new catalytic pathway.

It is important to note that although radiation damage is not fully prevented, its progression is significantly slowed down by freezing the crystal (by up to two orders of magnitude for X-rays [42,43]). Yet, the structure of the biological molecule in the crystal can be perturbed during the flash-freezing procedure [44]. The usually necessary addition of cryo-protectants to the crystal may also have an effect on the biological molecule's structure.

5 Data processing and model refinement

This section will describe the specificities of neutron crystallography from the processing of diffraction patterns to the refinement of molecular models. The goal here is not to describe every step in full detail, as these are very similar to X-ray crystallography (already well documented), but to focus on how crystallographic data treatment pertains to neutron crystallography.

5.1 Data reduction

For any crystallography experiment, processing of all recorded diffraction patterns yields, for each measured point of the reciprocal lattice (of Miller index *hkl*), an averaged intensity ($<I>$) and its corresponding error ($\sigma(I)$), which are then modified into an averaged structure factor amplitude ($<F>$) and its corresponding error ($\sigma(F)$). Although often appropriately scaled, the structure factor amplitudes are the square root of the measured intensities. Such data reduction requires a precise description of both the crystalline system (unit cell parameters and symmetry) and the instrument (detector geometry, crystal-detector distance, wavelength spectrum, *etc*...). Depending on the neutron source and the diffraction technique used for a given neutron protein diffractometer, different processing packages have been developed.

For monochromatic instruments, traditional software developed for X-ray crystallography can be directly used or slight modifications are needed. For example, the Biodiff instrument readily uses the hkl2000 software [45], while XDS [46] is being adapted for the D19 diffractometer of ILL. Quasi-Laue instruments generally utilize the *LAUEGEN* suite, which was originally developed to reduce X-ray Laue diffraction data [47]. The particularity of the quasi-Laue neutron diffraction data is their propensity to be overlapped. Dealing with such is non-trivial as the data reduction software needs to extract a single set of ($\langle I \rangle$, $\sigma(I)$) from two or more overlapped reflections. Another level of complexity arises from the multitude of wavelengths which can stimulate the same reciprocal lattice point with different crystal orientations. The recorded diffracted intensities have to be normalized with respect to the incident stimulating wavelength. Two factors come into play here: i) the diffracted intensity is proportional to λ^4(see Eq. (4) above) and ii) the incident neutron flux is dependent of the wavelength. This wavelength normalization procedure and the scaling of all reflections is performed with the software *LSCALE* [48]. In the quasi-Laue method, a proper estimation of the individual intensities of overlapped reflections is extremely difficult, and in this last step of data reduction, multiple intensity measurements of the same *hkl* index from different diffraction patterns or from symmetry equivalent reflections are compared, and, if they deviate too much, the software is unable to provide an accurate estimation of the intensity so those reflections are discarded from the data. This is a limitation of the quasi-Laue technique. The overall completeness is generally about ~80% and ~50% in the highest resolution shell, due to the number of overlapped reflections increasing with resolution. The unit-cell dimensions are therefore limited to ~150Å for quasi-Laue techniques with a wavelength spread of about 30%. In this perspective, the new protein diffractometer of the ILL has been designed with a wavelength spread of 10% to extend the capabilities of neutron protein crystallography at ILL.

For TOF Laue diffractometers at spallation sources, specific data reduction software have been developed to cope with the peculiar spatial arrangement of their array of detectors. At MaNDi, 45 detectors are positioned in a spherical way around the sample to maximize the reciprocal space coverage. Software developments is still ongoing with new procedures to integrate peaks from these diffractometers via neural networks for MaNDi [49], or the addition of profile fitting at iBIX [50].

The resolution of the diffraction dataset is therefore based on several data quality indicators [51]. The R_{merge}, and subsequently the R_{pim}, have conventionally provided an estimate of the precision of the individual measurements. The difference between the two is that the latter is weighted to take into account the redundancy in the data. The R_{merge} drastically rises with increased data multiplicity, and therefore does not provide accurate insights into data quality. Data measured multiple times should become more accurate.

The I/σ ratio provides another indication about which scattering data are reasonable to include in the dataset. However, the previous strict I/ σ ~2 ratio which defined the highest resolution has been relaxed, so as to include as much data as possible in the dataset (I/σ >1) (detectors are much more accurate nowadays…). A more rigorous statistical approach is to use the correlation coefficient between half data sets. Typically, $CC_{1/2}$ is ~ 1.0 at low resolution, and drops toward 0 with increasing resolution, but also decreasing signal-to-noise ratio. The resolution with $CC_{1/2}$ cut-offs between 0.2 and 0.5 have been found to be acceptable.

5.2 The phase problem

While intensities are measured in diffraction experiments, phases, needed to compute the electron and nuclear densities in X-ray and neutron crystallography respectively, have to be retrieved. Phases can be experimentally obtained (a good introduction to the problem and its solution(s) in [52] and references therein) or retrieved from a previously obtained homologous model [53] (homologous here refers to the protein scaffold being similar). While the experimental phasing in protein crystallography is exclusively performed with X-ray, a proof of concept has been performed at ILL [54]. For neutron crystallography, experimenters already have a model from X-rays, which can be used to obtain the phases. Hydrogens or deuterium are then added to the model in their riding positions, based only on the stereochemistry. They will further be refined using the neutron diffraction data.

5.3 The molecular model; the pdb format

The aim of crystallography is to produce a 3-dimensional atomic model of the repeating asymmetric unit of the biological crystal. This asymmetric unit may contain one or more molecules.

The Protein Data Bank archive (PDB, http://www.rcsb.org) serves as a world-wide repository of information about the structures of biological macromolecules, like protein and nucleic acids. High-resolution structures were determined by X-ray and neutron crystallography, electron microscopy and NMR. The repository also includes pseudo-atomic models derived from lower resolution structural techniques. Importantly, prior to publication, coordinates, and associated experimental information (including diffraction structure factors, and NMR constraints) must now be deposited to the PDB. The entries are therefore validated and annotated following a common set of criteria and accessible to all [55].

The PDB file is in a plain-text format, which not only contains the atomic coordinates but also holds a variety of extra information regarding the biomolecule itself, and the crystal if the model has been derived from a crystallography experiment. The first section of the PDB file contains a series of remarks, giving information about the protein and ligands, the quality of the model and the crystal characteristics. The individual atoms in the model are subsequently defined by four variables: their position in 3-dimensional space (x,y,z), and the temperature factor (or B-factor). The later value provides an error estimate in the atom's position (at high resolution, the B-factor can be separated out into 3-dimensions too).

The accuracy of the model is obviously limited by the quality of the experimental data used to build the model. For example, at low resolution, multiple chemically correct models can be built that fit the available experimental data. Atoms that are not accounted for by the diffraction data are typically omitted from the model, or placed in chemical acceptable positions, but with zero occupancy. In X-ray structures, due to resolution, the model therefore does not often list hydrogen positions, while in neutron structures, observed hydrogens and deuteriums are detailed.

The deposited structure is the best model that fits the diffraction data and the chemistry. A priori biological knowledge may also have provided a guiding hand in model building. The precision that atoms can be positioned in the model depend on the regularity of the crystal [56], and a well-ordered crystal will likely yield atomic positions with better precision.

A structure determined to 1-Angstrom resolution would allow individual atoms to be directly positioned in the structural model. At resolutions better than 3-Angstrom, the shape of the amino acid can be clearly seen, but here, as resolution gets worse, the chemistry of the model plays an increasingly important role to limit the possible positions of the atoms can have in the structure. In particular, the ability to distinguish between oxygens, nitrogens and carbons is compromised, which raises issues with the orientation of Asparagine, Glutamate and Histidine side chains. In the case that diffraction is worse than 4-Angstrom resolution, secondary structures elements become difficult to identify. An interesting series of movies has been realized by James Holton (https://bl831.als.lbl.gov/~jamesh/movies/) which stress out the importance of various parameters of dataset on the resulting density maps, in which the crystallographer has to build his molecular model (here done with X-rays, so electron density).

5.4 Model refinement and validation

In crystallography, the model refinement is a computationally intensive procedure which aims at minimizing the differences between the observed structure factors (derived from the experimental intensities) and the calculated ones (computed from the atomistic pdb model). This minimization is done using different strategies, like the refinement of atom positions in space (x,y and z), as well as the atomic displacement parameter (ADP or B factor), which is linked to the thermal agitation of an atom. Several statistical tools (including the R_{factor} and R_{free}) have been introduced as metrics to follow the evolution of the refinement, and, as in any multi-parametric problem, making sure that over-fitting is not arising. Again, for more information regarding these particular concepts, Kleywegt and Jones [57] is an interesting and complete reference.

At moderate resolution (2.3-2.5Å), each atom in a structure is typically described by four parameters (x,y,z and B factor). When compared to X-rays, neutron refinement raises the level of complexity. The addition of H/D atoms in the refinement procedure will double the number of atoms in the structure and consequently the number of parameters to refine for an identical number of experimental data (Bragg peaks). When refining a neutron structure, the data to parameter ratio is therefore low (even below 1) and the problem is underdetermined. However, chemical restraints can be used to increase the observations over parameter ratio, which makes neutron refinement possible on its own. An alternative strategy is to perform refinement using both neutron and X-ray experimental data against the same molecular model. *Phenix* propose such strategies [58], *SHELXL2013* [59] has been modified to easily perform neutron refinement and a neutron version of *Refmac* [60] is under development.

6 Neutron crystallography serves the chemistry of biological macromolecules

The structural details derived from the neutron scattering of H/D atoms of the macromolecule of interest is highly valuable in the understanding of various aspects of its biological function [61,62]. Since all biological reactions occur in a watery environment, hydrogen bonds are therefore ubiquitous in biology. Not only they hold the basic structure of proteins (from secondary to tertiary structure) but they coordinate solvent interaction, ligand binding, recognition and transformation. Many biological processes require the transfer of a hydrogen atom from the protein to the substrate or *vice-versa*. Hydrogen bonds are directional and formed between an electronegative donor, which shares its covalently bound H atom with another electronegative acceptor atom. The corresponding energy of such bond ranges between 2 and 10 kcal.mol^{-1}, depending on the geometry between the donor, acceptor and the shared H atom (distances and angles) [63]. This relatively low energy allows, for example, the H bonds to be formed and broken for substrate recognition and product release, respectively, over a large range of temperatures and solvent conditions (pH in particular). Importantly, although the position of most hydrogens bound to carbons can be accurately predicted, this is not the case for hydrogens attached to hydrogen-bond donor atoms like oxygen, sulfur and even nitrogen atoms. Unlike what had been observed in small molecules, very few truly colinear hydrogen bonds are observed in proteins, due to the steric constraints imposed by atom packing. Of course, the presence of a given hydrogen-bond can depend on the protonation state of charged amino-acids and their identification using neutron

 https://doi.org/10.1051/epjconf/202023602001

crystallography is generally essential to decipher the biological function. Having access to the hydrogen bond network in proteins permits the visualization of ligand/inhibitor binding modes or of specific protonation states of key catalytic residues, the clear identification of the hydration pattern of the catalytic site, as well as the observation of intermediates steps along the catalytic reaction. All these aspects will be stressed-out in this section, with precise and concrete examples of recent neutron diffraction studies.

6.1 Canonical and unusual hydrogen bonds

While neutron protein crystallography is unique to identify the position of the hydrogen atoms at near-atomic resolution and their implication in H-bonds, the technique also helped to discover some atypical ones. One of them is the low-barrier hydrogen bond (LBHB), first hypothesized in the early 90s [64], in which the hydrogen atom is literally shared between the donor and acceptor atoms (*i.e* at equal distance). Such atomistic arrangement is possible if the pKa values of donor and acceptor atoms are matched. The unusual nature of this hydrogen bond has been repeatedly observed in several neutron protein structures, such as, citrate synthase [65], serine protease [66], ketosteroid isomerase [67] and the fluorescent PYP protein [68–70]. It has also been discovered in the catalytic triad of the aminoglycoside-*N3*-acetyltransferase VIa (AAC-VIa) with a putative role in enzyme catalysis. In a follow up study, a structural proof has been obtained on the catalytic potential of such LBHB. Indeed, in the neutron crystal structure of the protein bound to its two most efficient substrates (*i.e* gentamicin and sisomicin, and in absence of cofactor to prevent the reaction to occur), a LBHB is observed within the catalytic triad (Figure 5a). On the contrary, a canonical hydrogen bond is observed for one of its least preferred substrates (kanamycin B, Figure 5b). This study provided the first structural evidence of the importance of a LBHB in catalytic efficiency, and brought to a close a long-lived debate on the potential role of these atypical hydrogen bonds.

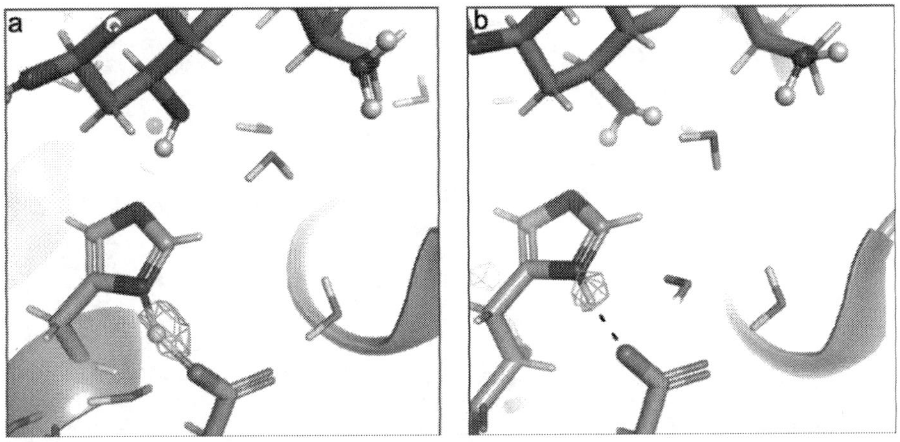

Fig. 5. Atypical and canonical hydrogen bonds. Structure of the catalytic site of the aminoglycoside-N3-acetyltransferase VIa in the presence of *(a)* gentamicin containing low-barrier hydrogen bonds, and *(b)* kanamycin B, having canonical hydrogen bonds. Gentamicin and kanamycin B are displayed as marine and magenta sticks, respectively. mFo-DFc maps have been computed with the DD1 atoms of His 189 removed from the model. Positive nuclear density from this map is shown at 2.9σ in green. The shared deuterium in panel *(a)* is shown as a sphere, as well as all deuterium atoms of the antibiotics.

6.2 Ligand or inhibitor recognition

H-bonds are key contributors in ligand or inhibitor recognition. While most of the molecular models of such complexes are obtained with X-ray crystallography, the hydrogen bond network can only be inferred. Neutron crystallography remains a unique technique to unravel the exact and precise hydrogen bond network which holds the small molecule in the targeted binding site. In recent years, several neutron crystal structures of proteins bound to clinical drugs have been reported.

Farnesyl pyrophosphatase synthase is a molecular target of osteoporosis [71] which has recently been solved bound to the clinical drug risedronate by neutron crystallography [72]. This structure provides insight about the binding mode of risedronate and therefore insights for a rational improvement to the drug.

The aspartyl protease of HIV-1 is essential for the virus life-cycle, and non-infectious viral particles are produced if this protease is not properly functioning. Antiviral inhibitors of the protease are already marketed for AIDS treatment. The neutron structure of HIV-1 protease was solved bound to the clinical drug amprenavir and pictured an hydrogen bond network very different from the one inferred from the X-ray crystal structure [10] (Figure 6). Again, these structural details are valuable for the design of inhibitors of improved efficiency, especially due to the rapid apparition of drug resistance in the HIV, linked to its protease. Subsequently, a structure of an HIV-1 protease triple mutant, which was resistant to amprenavir, was surprisingly obtained in complex to the drug. Intriguingly, the room temperature neutron structure exhibited differences with the cryo-cooled X-ray structure of the same mutant. These data clearly demonstrated that the mutations did not alter the binding mode of amprenavir to the protein, but that other effects were at play to create drug resistance, such as, possibly global protein dynamics and active site conformational flexibility.

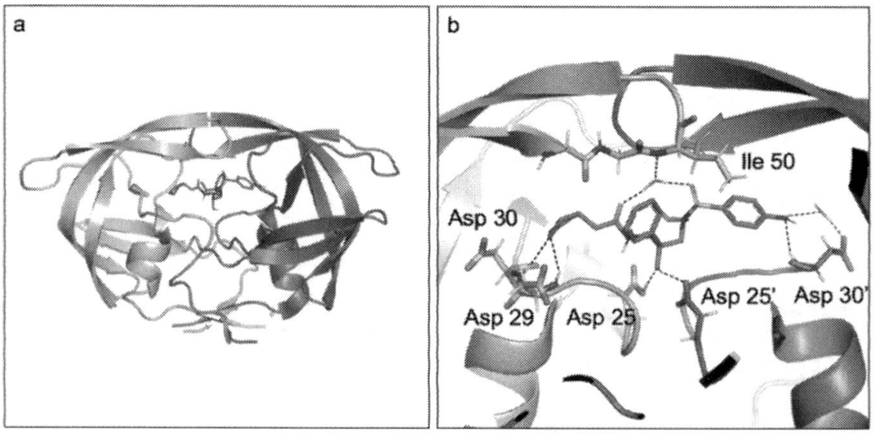

Fig. 6. Binding mode of amprenavir onto the HIV-1 protease. (a) Overall cartoon view with each monomer colored in grey and marine; the drug amprenavir is shown as magenta sticks. (b) Close up view with the important residues of the protease represented as sticks. For clarity, hydrogen atoms of the drug have not been displayed. H-bonds are shown as black dashes.

Another example of clinical drugs bound to proteins and obtained by neutron crystallography is the human carbonic anhydrase bound to either acetazolamide [11] or to methazolamine [12]. This enzyme is inhibited during the therapeutic treatment of glaucoma and epilepsy.

6.3 Protonation states

The electrically charged state of a single side chain can play a significant role in the function of the protein. The associated pKa is dependent on the side chain's local environment, and, in an enzyme's active site, it is finely tuned to match the protonation state required for the catalytic reaction to proceed. Neutron crystallographic studies were essential in identifying active site residues with unusual pKa's and therefore unusual protonation states. For example, the catalytic function of the aspartyl protease HIV-1 was studied at various pHs. This homodimeric protease bears a pair of aspartic acid residues involved in the peptide bond cleavage. Each of these two aspartates belong to a monomer and the active site lies at the dimer interface (Figure 6a). The most interesting feature is that the chemical nature of these two, supposedly negatively charged residues and within interacting distance from each other, is different. In the neutron structure bound to the clinical drug amprenavir (pH=6.0), the aspartic residue from monomer A is neutral (with a D atom covalently linked to $O\delta 1$) while the other one is negatively charged (Figure 7a). Interestingly, the same protein in complex with the drug darunavir reveals that the charge is on the aspartic acid of monomer A instead. The other aspartic residue shares a deuterium atom with the drug *via* a low-barrier hydrogen bond (Figure 7b). The position of the oxygen atoms in the carboxylate groups of these two aspartic residues are equivalent in the two structures, and that the deduction of proton positions and hydrogen bonding from distances between oxygens is at risk. Such subtle changes in the protonation states of catalytic residues are only accessible with neutron crystallography.

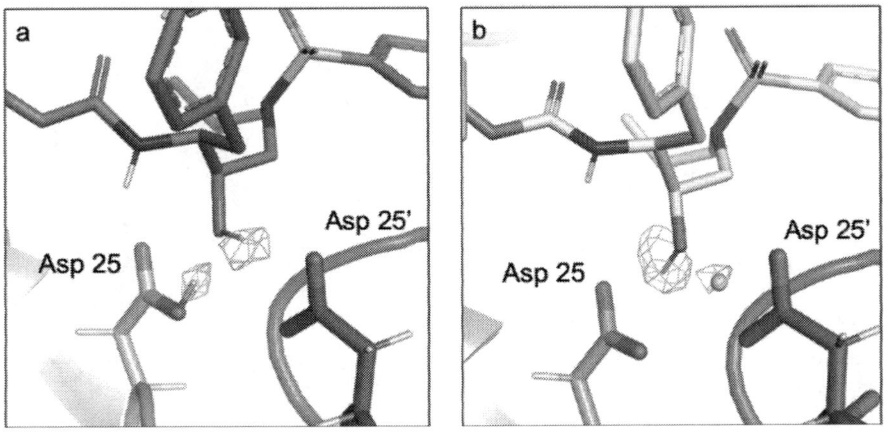

Fig. 7. Changes in protonation states of the aspartyl protease HIV-1 catalytic triad upon binding of different drugs. *(a)*, amprenavir (magenta sticks) and *(b)* duranavir (yellow sticks). mFo-DFc have been computed without the D atoms around the catalytic aspartate and nuclear density is shown at 4 and 3σ for *(a)* and *(b)*, respectively.

Protonation states of key residues in the chromophore pocket of fluorescent proteins were assessed with neutron crystallography, such as the chromoprotein Dathail [73] and the photoactive yellow protein (PYP) [70]. The protonation state of the chromophore itself has been observed, as well as those of surrounding residues, with an unexpected neutral arginine in the PYP chromophore pocket.

6.4 Solvent network

Heavy water (D_2O) has a characteristic-boomerang shaped nuclear density, as its three atoms have positive and equivalent coherent neutron scattering cross-sections. However, this typical shape is observed in structures of relatively high resolution (1.5Å and higher). A combination of X-ray (to locate the oxygen atom) and neutron (to position the deuterium atoms) diffraction make it possible to orient water molecules at much more moderate resolution (between 2.0 and 2.5Å, depending on the data quality)

For example, in the structure of crambin, a small protein of 46 residues, neutron diffraction data from an H/D exchanged crystal have been obtained at 1.1Å resolution [74]. The electron and nuclear density maps of five water molecules of this structure are shown in Figure 8 and reveal the level of details which can be obtained with neutron crystallography regarding the solvent network. Of course, not all neutron structures are solved at such high resolution, but the water orientation can be retrieved in moderate resolution structure by combining the information from X-ray (to first position the oxygen atom) and neutron (to then add the deuterium atoms in the nuclear density map). Water networks are also important for proton transfer, as observed in the human carbonic anhydrase II (HCA II) enzyme. The proton transfer is the final step of the catalytic reaction which involves a water molecule, part of a well-ordered 8Å thick water network that spans from the catalytic site to the bulk solvent [16,17].

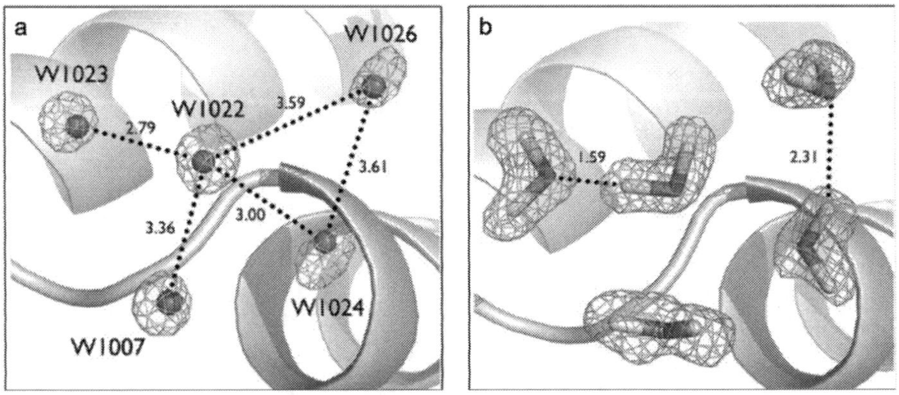

Fig. 8. 2mFo-DFc electron *(a)* and nuclear *(b)* density map around five solvent molecules. From the electron density, a large number of hydrogen bonds are geometrically possible. The nuclear density map clearly identifies which ones are present. Reproduced from [74]

In some cases, prior to ligand association, the ligand-binding pocket has a specific hydration pattern, which needs to be accommodated upon ligand binding [15]. High resolution X-ray and neutron structure of trypsin were obtained in its free form and complexed with two inhibitors. Without ligand, the binding pocket is filled with water molecules (Figure 9a) of high mobility, and characterized by an incomplete H-bond network, which does not fully solvate catalytic residue Asp 189. Upon ligand binding, water molecules are displaced and Asp 189 is engaged in two hydrogen bonds (Figure 9b). The precise understanding of water configuration in the ligand-free state gives clue on the forces which drive ligand-binding, with imperfect hydration of Asp 189 being one of them.

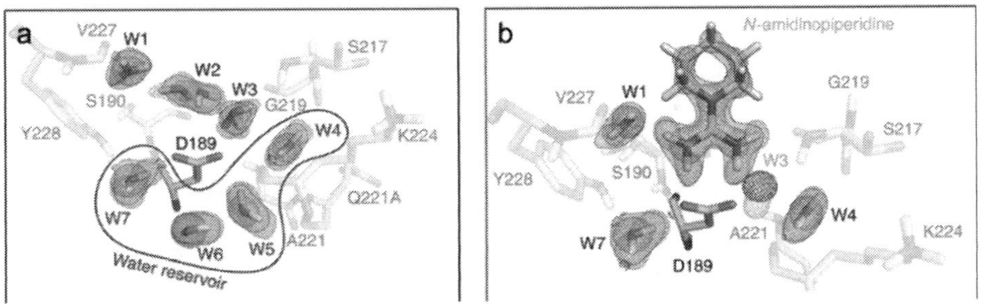

Fig. 9. Solvation of the uncomplexed trypsin (a) and bound to N-amidinopiperidine (b). Protein residues are depicted as light brown sticks. X-ray (orange mesh) and neutron 2mF$_o$-DF$_c$ (grey surface) maps are contoured at 1.5 and 1.8σ. The neutron omit mFo-DFc is represented in green (4σ) and was obtained after removal of deuterium atoms from all depicted water molecules as well as the deuterium atom of the inhibitor in *(b)*. The water reservoir made up by waters W4–W7 is highlighted in blue. Reproduced from [15].

6.5 Catalytic intermediates

Finally, neutrons provide the unique capability to decipher the enzymatic mechanisms when trapping catalytic intermediates. Several approaches are available, *via* substrate analogues which stop the reaction at a given stage [6] or the trapping of intermediates by flash-freezing [7,8]. Recently, a nice study reported two enzymatic intermediates within a single crystal. In the asymmetric unit contained, each of two monomers were trapped in two consecutive steps of the pathway. In the first one, the reaction fully processed to reach the stage at which the substrate analog prevents the reaction to further advance, while in the second one, the catalytic loop, involved in crystal contacts, could not close onto the substrate analog and the active site, and the reaction was therefore stopped a step before that of the other monomer. This study was performed on an aspartate aminotransferase, a pyridoxal 5′-phosphate (PLP) dependent enzyme. This cofactor, derived from pyridoxine is a ubiquitous cofactor in biology, needed for more that 140 different biochemical transformations. The neutron structure revealed clear protonation states of the catalytic pocket in the two complexes (Figure 10). The consequence of the fine undersnanding of the mechanism at the atomistic level is the re-thinking of the catalytic mechanism of such an important and widespread class of enzyme. While these studies are the most challenging for neutron crystallography, they usually are the most useful and informative.

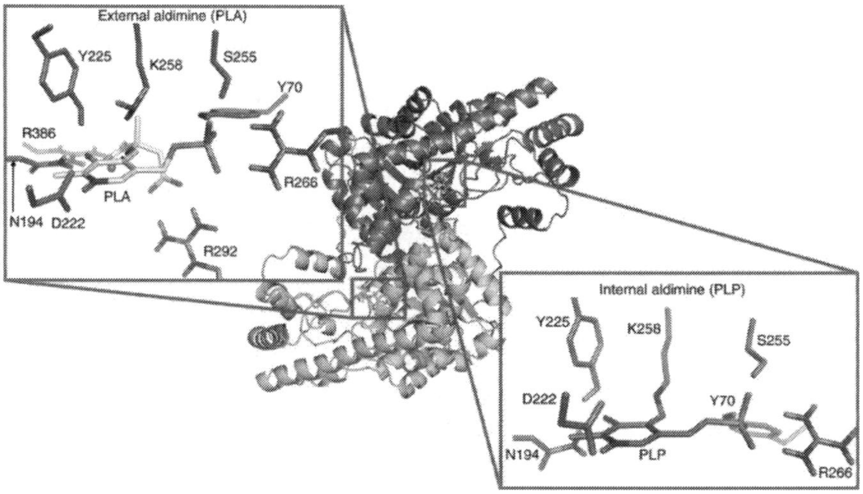

Fig. 10. Structure of aspartate aminotransferase with substrate analog α-methylaspartate. In monomer A (light green), the catalytic loop closes and the reaction reaches the so-called external aldimine (PLA). In monomer B (dark green), the catalytic loop remains open and the reaction is stopped at the internal aldimine step of the reaction. Cofactor is represented as grey (PLP, internal aldimine) or yellow (PLA, external aldimine) sticks. Key residues are shown in grey sticks, with deuterium in green.

References

[1] M.P. Blakeley, S.S. Hasnain, S.V. Antonyuk, IUCrJ 2 (2015) 464–474.
[2] J.C.-H. Chen, C.J. Unkefer, IUCrJ 4 (2017) 72–86.
[3] A.Y. Kovalevsky, L. Hanson, S.Z. Fisher, M. Mustyakimov, S.A. Mason, V.T. Forsyth, M.P. Blakeley, D.A. Keen, T. Wagner, H.L. Carrell, A.K. Katz, J.P. Glusker, P. Langan, Structure 18 (2010) 688–699.
[4] O. Gerlits, T. Wymore, A. Das, C.-H. Shen, J.M. Parks, J.C. Smith, K.L. Weiss, D.A. Keen, M.P. Blakeley, J.M. Louis, P. Langan, I.T. Weber, A. Kovalevsky, Angewandte Chemie 128 (2016) 5008–5011.
[5] O. Gerlits, D.A. Keen, M.P. Blakeley, J.M. Louis, I.T. Weber, A. Kovalevsky, J. Med. Chem. 60 (2017) 2018–2025.
[6] S. Dajnowicz, R.C. Johnston, J.M. Parks, M.P. Blakeley, D.A. Keen, K.L. Weiss, O. Gerlits, A. Kovalevsky, T.C. Mueser, Nat Commun 8 (2017) 955.
[7] C.M. Casadei, A. Gumiero, C.L. Metcalfe, E.J. Murphy, J. Basran, M.G. Concilio, S.C.M. Teixeira, T.E. Schrader, A.J. Fielding, A. Ostermann, M.P. Blakeley, E.L. Raven, P.C.E. Moody, Science 345 (2014) 193–197.
[8] H. Kwon, J. Basran, C.M. Casadei, A.J. Fielding, T.E. Schrader, A. Ostermann, J.M. Devos, P. Aller, M.P. Blakeley, P.C.E. Moody, E.L. Raven, Nat Commun 7 (2016) 13445.
[9] S.K. Panigrahi, G.R. Desiraju, Proteins 67 (2007) 128–141.
[10] I.T. Weber, M.J. Waltman, M. Mustyakimov, M.P. Blakeley, D.A. Keen, A.K. Ghosh, P. Langan, A.Y. Kovalevsky, J. Med. Chem. 56 (2013) 5631–5635.
[11] S.Z. Fisher, M. Aggarwal, A.Y. Kovalevsky, D.N. Silverman, R. McKenna, J. Am. Chem. Soc. 134 (2012) 14726–14729.
[12] M. Aggarwal, A.Y. Kovalevsky, H. Velazquez, S.Z. Fisher, J.C. Smith, R. McKenna, IUCrJ 3 (2016) 319–325.
[13] M.-M. Blum, F. Löhr, A. Richardt, H. Rüterjans, J.C.-H. Chen, J. Am. Chem. Soc. 128 (2006) 12750–12757.
[14] M.-M. Blum, M. Mustyakimov, H. Rüterjans, K. Kehe, B.P. Schoenborn, P. Langan, J.C.-H. Chen, PNAS 106 (2009) 713–718.
[15] J. Schiebel, R. Gaspari, T. Wulsdorf, K. Ngo, C. Sohn, T.E. Schrader, A. Cavalli, A. Ostermann, A. Heine, G. Klebe, Nature Communications 9 (2018).

[16] D.N. Silverman, R. McKenna, Acc. Chem. Res. 40 (2007) 669–675.
[17] Z. Fisher, A.Y. Kovalevsky, M. Mustyakimov, D.N. Silverman, R. McKenna, P. Langan, Biochemistry 50 (2011) 9421–9423.
[18] T.P. Halsted, K. Yamashita, C.C. Gopalasingam, R.T. Shenoy, K. Hirata, H. Ago, G. Ueno, M.P. Blakeley, R.R. Eady, S.V. Antonyuk, M. Yamamoto, S.S. Hasnain, IUCrJ 6 (2019) 761–772.
[19] D. Kekilli, F.S.N. Dworkowski, G. Pompidor, M.R. Fuchs, C.R. Andrew, S. Antonyuk, R.W. Strange, R.R. Eady, S.S. Hasnain, M.A. Hough, Acta Crystallogr. D Biol. Crystallogr. 70 (2014) 1289–1296.
[20] M. Suga, F. Akita, K. Hirata, G. Ueno, H. Murakami, Y. Nakajima, T. Shimizu, K. Yamashita, M. Yamamoto, H. Ago, J.-R. Shen, Nature 517 (2015) 99–103.
[21] U. Shmueli, Theories and Techniques of Crystal Structure Determination, Oxford University Press, Oxford, New York, 2007.
[22] C. Wilkinson, M.S. Lehmann, Nuclear Instruments and Methods in Physics Research Section A: Accelerators, Spectrometers, Detectors and Associated Equipment 310 (1991) 411–415.
[23] Z. Dauter, Acta Cryst D 55 (1999) 1703–1717.
[24] S. Brockhauser, R.B.G. Ravelli, A.A. McCarthy, Acta Cryst D 69 (2013) 1241–1251.
[25] M.P. Blakeley, S.C.M. Teixeira, I. Petit-Haertlein, I. Hazemann, A. Mitschler, M. Haertlein, E. Howard, A.D. Podjarny, Acta Cryst D 66 (2010) 1198–1205.
[26] G.C. Schröder, W.B. O'Dell, D. a. A. Myles, A. Kovalevsky, F. Meilleur, Acta Cryst D 74 (2018) 778–786.
[27] A. Ostermann, T. Schrader, Journal of Large-Scale Research Facilities JLSRF 1 (2015) 2.
[28] I. Tanaka, K. Kurihara, T. Chatake, N. Niimura, J Appl Cryst 35 (2002) 34–40.
[29] L. Coates, A.D. Stoica, C. Hoffmann, J. Richards, R. Cooper, J Appl Cryst 43 (2010) 570–577.
[30] I. Tanaka, K. Kusaka, T. Hosoya, N. Niimura, T. Ohhara, K. Kurihara, T. Yamada, Y. Ohnishi, K. Tomoyori, T. Yokoyama, Acta Crystallogr. D Biol. Crystallogr. 66 (2010) 1194–1197.
[31] K.H. Andersen, C.J. Carlile, J. Phys.: Conf. Ser. 746 (2016) 012030.
[32] N. Niimura, Y. Karasawa, I. Tanaka, J. Miyahara, K. Takahashi, H. Saito, S. Koizumi, M. Hidaka, Nuclear Instruments and Methods in Physics Research A 349 (1994) 521–525.
[33] F. Cipriani, J.C. Castagna, C. Wilkinson, M.S. Lehmann, G. Büldt, in: B.P. Schoenborn, R.B. Knott (Eds.), Neutrons in Biology, Springer US, Boston, MA, 1996, pp. 423–431.
[34] P. Langan, G. Greene, B.P. Schoenborn, J Appl Cryst 37 (2004) 24–31.
[35] L. Coates, M.J. Cuneo, M.J. Frost, J. He, K.L. Weiss, S.J. Tomanicek, H. McFeeters, V.G. Vandavasi, P. Langan, E.B. Iverson, J Appl Cryst 48 (2015) 1302–1306.
[36] L. Coates, L. Robertson, J Appl Crystallogr 50 (2017) 1174–1178.
[37] A. McPherson, J.A. Gavira, Acta Crystallogr F Struct Biol Commun 70 (2013) 2–20.
[38] N. Asherie, Methods 34 (2004) 266–272.
[39] N. Niimura, R. Bau, Acta Crystallogr., A, Found. Crystallogr. 64 (2008) 12–22.
[40] N. Junius, E. Oksanen, M. Terrien, C. Berzin, J.-L. Ferrer, M. Budayova-Spano, J Appl Crystallogr 49 (2016) 806–813.
[41] M.P. Blakeley, A.J. Kalb, J.R. Helliwell, D.A.A. Myles, (n.d.) 6.
[42] C. Nave, E.F. Garman, J Synchrotron Rad 12 (2005) 257–260.
[43] M. Warkentin, R.E. Thorne, Acta Cryst D 66 (2010) 1092–1100.
[44] J.S. Fraser, H. van den Bedem, A.J. Samelson, P.T. Lang, J.M. Holton, N. Echols, T. Alber, Proc. Natl. Acad. Sci. U.S.A. 108 (2011) 16247–16252.
[45] Z. Otwinowski, W. Minor, in: Methods in Enzymology, Academic Press, 1997, pp. 307–326.
[46] W. Kabsch, Acta Crystallographica Section D Biological Crystallography 66 (2010) 125–132.
[47] J.W. Campbell, Q. Hao, M.M. Harding, N.D. Nguti, C. Wilkinson, Journal of Applied Crystallography 31 (1998) 496–502.
[48] S. Arzt, J.W. Campbell, M.M. Harding, Q. Hao, J.R. Helliwell, J Appl Cryst 32 (1999) 554–562.
[49] B. Sullivan, R. Archibald, J. Azadmanesh, V.G. Vandavasi, P.S. Langan, L. Coates, V. Lynch, P. Langan, J Appl Cryst 52 (2019) 854–863.
[50] N. Yano, T. Yamada, T. Hosoya, T. Ohhara, I. Tanaka, K. Kusaka, Sci Rep 6 (2016) 36628.
[51] P.A. Karplus, K. Diederichs, Curr. Opin. Struct. Biol. 34 (2015) 60–68.
[52] G. Taylor, Acta Cryst D 59 (2003) 1881–1890.
[53] M.G. Rossmann, D.M. Blow, IUCr, Acta Crystallographica (1962).

[54] M.G. Cuypers, S.A. Mason, E. Mossou, M. Haertlein, V.T. Forsyth, E.P. Mitchell, Sci Rep 6 (2016).
[55] H. Berman, K. Henrick, H. Nakamura, Nat Struct Mol Biol 10 (2003) 980–980.
[56] J.P. Mackay, M.J. Landsberg, A.E. Whitten, C.S. Bond, Trends Biochem. Sci. 42 (2017) 155–167.
[57] G.J. Kleywegt, T. Alwyn Jones, in: Methods in Enzymology, Academic Press, 1997, pp. 208–230.
[58] P.D. Adams, P.V. Afonine, G. Bunkóczi, V.B. Chen, I.W. Davis, N. Echols, J.J. Headd, L.-W. Hung, G.J. Kapral, R.W. Grosse-Kunstleve, A.J. McCoy, N.W. Moriarty, R. Oeffner, R.J. Read, D.C. Richardson, J.S. Richardson, T.C. Terwilliger, P.H. Zwart, Acta Crystallographica Section D Biological Crystallography 66 (2010) 213–221.
[59] T. Gruene, H.W. Hahn, A.V. Luebben, F. Meilleur, G.M. Sheldrick, J Appl Cryst 47 (2014) 462–466.
[60] G.N. Murshudov, P. Skubák, A.A. Lebedev, N.S. Pannu, R.A. Steiner, R.A. Nicholls, M.D. Winn, F. Long, A.A. Vagin, Acta Crystallogr D Biol Crystallogr 67 (2011) 355–367.
[61] E. Oksanen, J.C.-H. Chen, S.Z. Fisher, Molecules 22 (2017) 596.
[62] W.B. O'Dell, A.M. Bodenheimer, F. Meilleur, Arch. Biochem. Biophys. 602 (2016) 48–60.
[63] T.W. Martin, Z.S. Derewenda, Nat. Struct Biol. 6 (1999) 403–406.
[64] J.A. Gerlt, P.G. Gassman, J. Am. Chem. Soc. 115 (1993) 11552–11568.
[65] Z. Gu, D.G. Drueckhammer, L. Kurz, K. Liu, D.P. Martin, A. McDermott, Biochemistry 38 (1999) 8022–8031.
[66] J.L. Markley, W.M. Westler, Biochemistry 35 (1996) 11092–11097.
[67] Q. Zhao, C. Abeygunawardana, A.G. Gittis, A.S. Mildvan, Biochemistry 36 (1997) 14616–14626.
[68] S. Anderson, S. Crosson, K. Moffat, Acta Cryst D 60 (2004) 1008–1016.
[69] S.Z. Fisher, S. Anderson, R. Henning, K. Moffat, P. Langan, P. Thiyagarajan, A.J. Schultz, Acta Cryst D 63 (2007) 1178–1184.
[70] S. Yamaguchi, H. Kamikubo, K. Kurihara, R. Kuroki, N. Niimura, N. Shimizu, Y. Yamazaki, M. Kataoka, Proceedings of the National Academy of Sciences 106 (2009) 440–444.
[71] P.D. Delmas, The Lancet 359 (2002) 2018–2026.
[72] T. Yokoyama, M. Mizuguchi, A. Ostermann, K. Kusaka, N. Niimura, T.E. Schrader, I. Tanaka, J. Med. Chem. 58 (2015) 7549–7556.
[73] P.S. Langan, D.W. Close, L. Coates, R.C. Rocha, K. Ghosh, C. Kiss, G. Waldo, J. Freyer, A. Kovalevsky, A.R.M. Bradbury, Journal of Molecular Biology 428 (2016) 1776–1789.
[74] J.C.-H. Chen, B.L. Hanson, S.Z. Fisher, P. Langan, A.Y. Kovalevsky, PNAS 109 (2012) 15301–15306.

EPJ Web of Conferences **236**, 03001 (2020) https://doi.org/10.1051/epjconf/202023603001
JDN 24

The basics of small-angle neutron scattering (SANS for new users of structural biology)

Cy M. Jeffries[1,*], *Zuzanna* Pietras[2], and *Dmitri I.* Svergun[1]

[1]European Molecular Biology Laboratory (EMBL), Hamburg Unit, c/o Deutsches Elektronen-Synchrotron, 22607 Hamburg, Germany

[2]Department of Physics, Chemistry, and Biology, Linköping University, 581 83 Linköping, Sweden

Abstract. Small-angle neutron scattering (SANS) provides a means to probe the time-preserved structural state(s) of bio-macromolecules in solution. As such, SANS affords the opportunity to assess the redistribution of mass, i.e., changes in conformation, which occur when macromolecules interact to form higher-order assemblies and to evaluate the structure and disposition of components within such systems. As a technique, SANS offers scope for 'out of the box thinking', from simply investigating the structures of macromolecules and their complexes through to where structural biology interfaces with soft-matter and nanotechnology. All of this simply rests on the way neutrons interact and scatter from atoms (largely hydrogens) and how this interaction differs from the scattering of neutrons from the nuclei of other 'biological isotopes'. The following chapter describes the basics of neutron scattering for new users of structural biology in context of the neutron/hydrogen interaction and how this can be exploited to interrogate the structures of macromolecules, their complexes and nano-conjugates in solution.

1 Introduction

Structural biologists have a vast array of techniques at their disposal to study the structures of proteins and other bio-macromolecules. The recent 'resolution revolution' in cryo-electron microscopy [1-4] has made a significant impact in the field while X-ray crystallography remains as the key method for determining atomic-level structural information [5]. Nuclear magnetic resonance spectroscopy continues to advance [6], with ever-more powerful instruments. In this arguably daunting environment, small-angle neutron scattering (SANS) affords an additional means to determine the low-resolution structures and dispositions of single- and multi-component bio-macromolecular systems [7-16]. Unlike cryo-EM (that, by definition, is performed in a frozen state), or X-ray crystallography (that requires a diffraction-quality crystal), samples for SANS experiments are typically prepared in aqueous buffers [17], although there is no restriction on sample-type or environment. Gels [18, 19], conjugates [20], phases [21]; trapped/confined/emulsion states [22, 23]; shear or pressure states [24] are all amenable to SANS. The reason for this diversity is due to the way

* Corresponding author: cy.jeffries@embl-hamburg.de

in which neutrons interact with matter: Neutrons scatter predominantly from atomic nuclei and do so depending on the isotope-specific neutron-nuclei interaction. As such, neutrons are both deeply-penetrating and, unlike X-rays that are scattered or absorbed from electrons, do not directly affect electronic transitions, i.e., are relatively chemically inert (although their absorption into materials and the production of secondary α- β and γ radiation over time will result in deleterious biological effects). Additional physical parameters of the neutron are summarized in Figure 1.

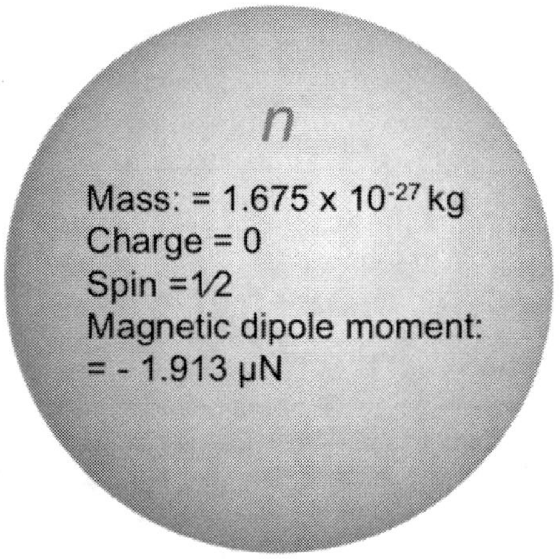

Fig. 1. *General properties of the neutron.*

Neutrons have a mass, zero charge, a spin-state of ½ and a magnetic dipole moment and may be described as particles or as waves with a wavelength λ, i.e., display 'wave-particle duality'. The magnetic dipole moment may be utilized for magnetic neutron scattering, although this is rarely employed in biological investigations. For most structural biology experiments, SANS focuses on measuring the elastic (i.e. without energy transfer) and coherent (i.e. direction-dependent) scattering of neutrons (Figure 2) from the atomic nuclei of the 'biological isotopes' found in proteins, polynucleotides, carbohydrates, lipids, etc.

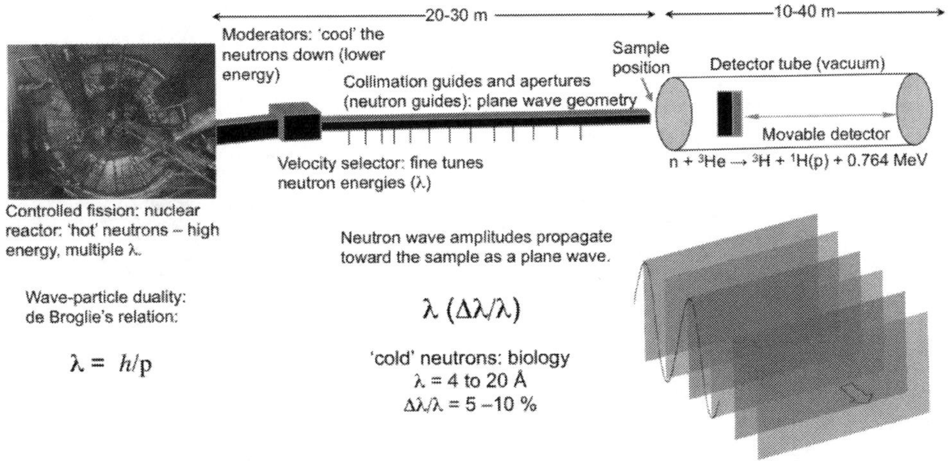

Fig. 2. *Basic schematic of a SANS instrument*. A neutron source, e.g., a reactor, produces neutrons with multiple energies. The neutrons have wave-particle duality, as described by the de Broglie's formula that relates the wavelength of a particle to the ratio of Planck's constant, h, and the momentum, p. Therefore, the neutrons from the source have high-momentums, short wavelengths, and high energy distribution. On exiting the reactor chamber, moderators (e.g., liquid helium) cool the neutrons down, i.e., slow them down to reduce the energy, with an increase in the corresponding wavelength. However, these 'cold neutrons' still have wide wavelength distribution. For standard SANS, it is often desired to illuminate a sample with a neutron that has a defined wavelength, λ, within a narrow bandwidth (Δλ/λ) due to the inverse relationship between λ and the momentum transfer of the scattered neutrons. A velocity selector, for example a turbine-like fan, spins at a particular rate to allow only those neutrons with the selected energy, and corresponding wavelength, down the neutron guides toward the sample. A set of guides and apertures (or 'virtual point sources') collimate the neutrons such that they can be visualized as propagating toward the sample as a plane-wave (or alternatively, from each aperture/virtual source, can be viewed as a spherical-wave with infinite radius). Upon interacting with the sample, some neutrons undergo scattering events, where there is momentum transfer, and propagate toward a detector that is often comprised of elements with very large neutron absorption cross sections, e.g., the isotope helium-3. The ^{3}He nuclei of the detector absorb the scattered neutron to generate tritium and a proton. At the proton is charged, its signal can be converted into a voltage/current, thereby counting the number of otherwise-neutrally-charged scattered neutrons.

The way each isotope's contribution to the summed scattering event proceeds (for all atoms in a sample, solution, and sample holder included) gives rise to what is, in effect, an interference pattern which is physically measured on a detector as intensities, I, as a function of angle, s, otherwise known as the momentum transfer. For the purposes of this chapter, the momentum transfer is defined as: $s = 4\pi\sin\theta/\lambda$ where 2θ is the scattering angle and λ the neutron wavelength, in keeping with the definition used for biological small-angle scattering [54]. (As an aside, the use of the letter s in this chapter is interchangeable with several other letter annotations found in the literature: $s=q=Q=k=h=\mu$ and should not be confused with the alternate scattering vector definition of S, where $S = 2\sin\theta/\lambda$). The SANS interference pattern (Figure 3) caused by time-preserved distance correlations between atomic nuclei within the volume boundary of a macromolecule, after the scattering from the solution/background has been subtracted, reflects the distance distribution of the scattering pairs inside the molecule and hence its overall structure. The great advantage of SANS is that it is possible to alter the interference pattern experimentally by altering the isotopic composition of the sample and/or the solvent, from which different types of structural information can be extracted. This information includes assessing fundamental structural parameters such as the radius of gyration, R_g [25, 26], and maximum particle dimension, D_{max}, as well as the probable frequency of scattering-pair distances in real space, or $p(r)$ [27, 28], that may be used to build low-resolution models of macromolecules, their complexes and higher-order assemblies[29, 30]. All of this largely depends on the unique way neutrons scatter from the most abundant isotope in the biological/universal sphere, regular light hydrogen, ^1H [15].

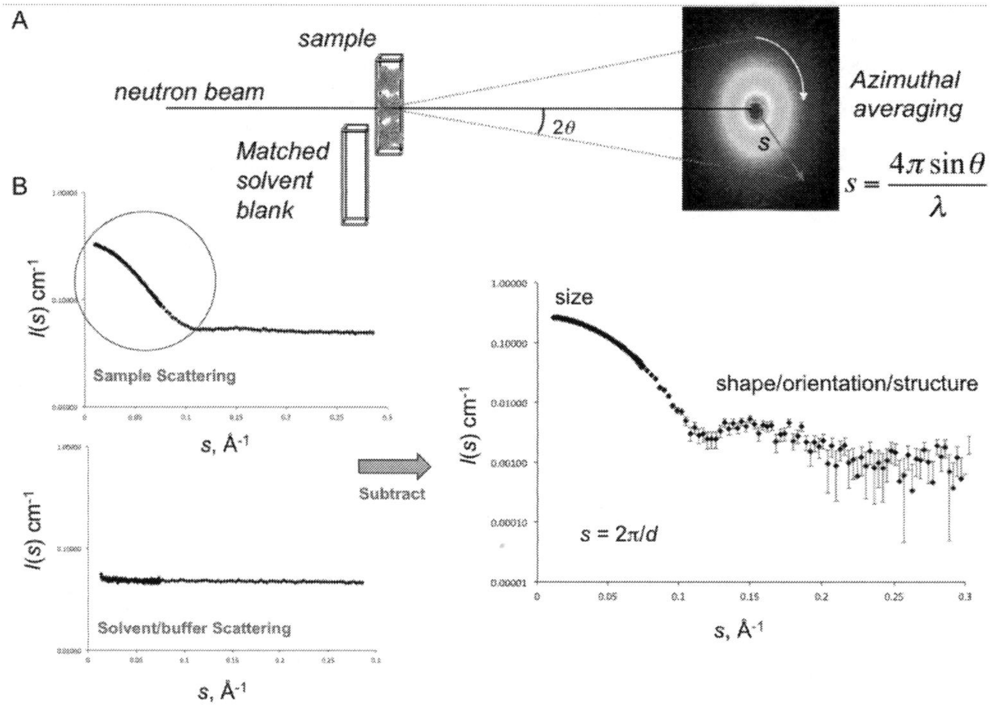

Fig. 3. *SANS data measurement and reduction*. **A.** A neutron hits a dilute solution of isotopically tumbling macromolecules in solution. In very rare instances, momentum is transferred to the neutron resulting in a scattering event at a particular scattering angle 2θ. The intensity of the scattering is recorded on a 2D-detector vs. the momentum transfer-vector with a modulus s, in units of inverse-distance (i.e., the scattering is recorded in momentum- or 'reciprocal-space'). The intensities recorded for each value of s are azimuthally averaged (a.k.a. radial averaging) to reduce the 2D-data to 1D-data, i.e., the neutron scattering intensities are plotted as $I(s)$ vs. s. **B.** The sample scattering consists of scattering contributions from the macromolecules in the sample *plus* neutron scattering contributions from the solvent. Preserved spatial correlations within the volume boundary of the macromolecules of the sample, give rise to a summed-coherent SANS signal 'over-and-above' the solvent scattering. The buffer does scatter, but for the distances probed in the SANS regime there are no preserved long-distance spatial correlations in the solvent, yielding 'flat scattering'. Solvent scattering contributions are subtracted from the sample scattering after the measurement of an exactly-matched solvent blank. After background subtraction and transmission corrections are applied (to take into account any differences in neutron absorption between the sample and the solvent) what remains are the coherent scattering intensities derived from the population of macromolecules in the sample. The intensity of coherent scattering signal encodes the frequency of preserved spatial correlations between scattering centres internal to the macromolecules, i.e., the size, shape and structure.

2 Neutrons

The detailed physics and mathematics underpinning neutron scattering are interesting but generally impenetrable for the non-physicist [31]. Since the discovery of the neutron by

James Chadwick in 1932 [32], and the subsequent pioneering work of Lise Meitner, Otto Frisch, Otto Hahn, and Fritz Straßmann [33, 34], nearly a century of highly dedicated theoretical and experimental research has taken place to describe what a neutron is and what it does when interacting with a nucleus. For the purposes of this chapter, the descriptions are overly-simplified and are provided to illustrate link between the underlying principles of neutron scattering and how it applies to structural biology.

There are three main outcomes when illuminating a sample with a neutron: 1), Nothing happens, i.e., the neutron travels through the sample and is transmitted; 2) The neutron may be absorbed by a nucleus that may result in secondary fissile events and the production of both ionizing and non-ionizing radiation and 3) Scatter (Figure 4). Neutron scattering events can be further sub-divided into:

i) Elastic scattering, where the neutron does not loose energy after scattering from a nucleus;

ii) Inelastic scattering, where the neutron changes energy after the scattering event;

iii) Nuclear-potential scattering where the scattered neutron from the sample relates to the spatial correlations between nuclei within the sample. If the distances between nuclei are preserved over time, this gives rise to a coherent SANS signal;

iv) Nuclear-spin scattering, where the scattered neutron does not necessarily relate to distance correlations between atom-pairs per-se, but relates to nuclear-spin orientations/correlations and internal-structural/time-dynamics of nuclear spin-transitions.

v) Combinations of the above and;

vi) 'Magnetic scattering' from unpaired-electron or exotic-magnetic lattice systems.

The focus of this chapter is points i) and iii), that is, utilizing the *elastic-coherent neutron scattering component* from which structure of macromolecules in solution or the disposition of macromolecular components within fully-formed complexes can be investigated.

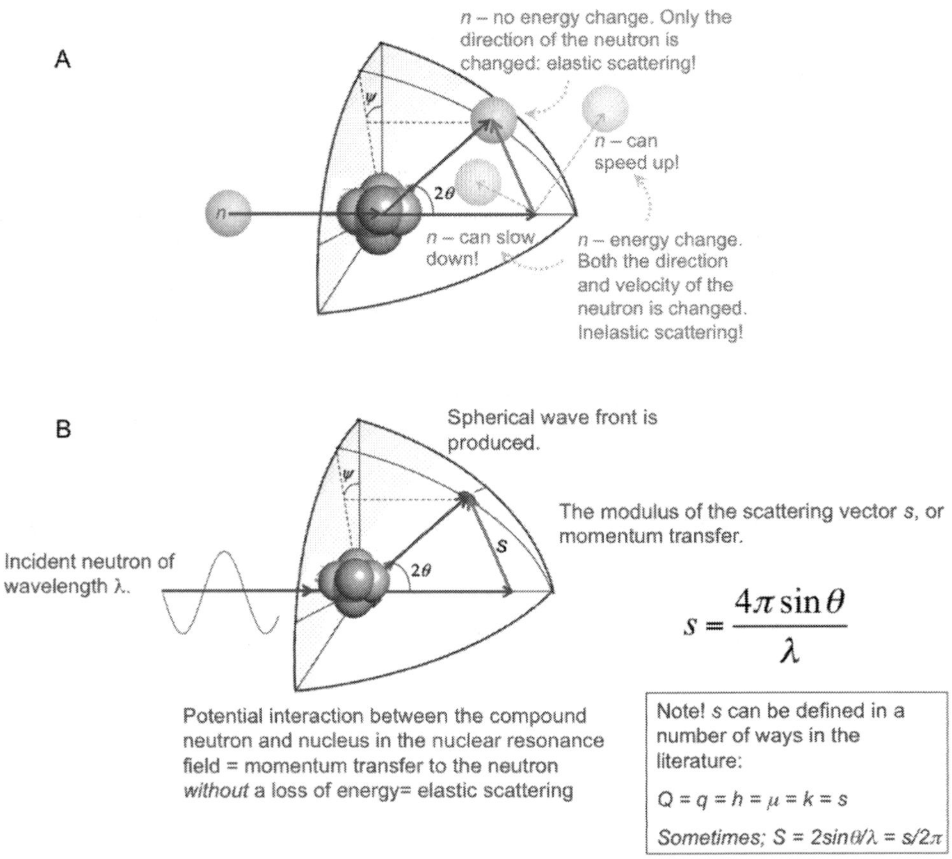

Fig. 4. *Neutron scattering interactions with a nucleus*. **A**. Neutrons can scatter from a nucleus, i.e., the transfer of momentum, such that they can speed up, slow down, and change direction. Elastic neutron scattering events yield in a change in angular momentum, without a loss in energy, i.e., a change in direction. Inelastic scattering events result in both a change in direction and speed of the neutron. **B.** SANS for structural biology relies on measuring the elastic scattering events arising from a nucleus-neutron interaction. When the incident neutron is viewed as a wave, the probability that a neutron will scatter at any point in any direction in any given time can, when viewed over all possible scattering angles 2θ, be mapped in terms of the wave-amplitude into a space that traces out a 'spherical wave front' where 2θ is described in terms of any point on this spherical wave by the momentum transfer, s.

2.1 Neutrons as waves…that can flip

It is perhaps easiest to understand of what a SANS pattern encodes if neutrons are conceived as waves. As with all waves, neutrons have a wavelength (that is inversely proportional to their energy), an amplitude, and a phase. In addition, the nucleus of each individual isotope has a probability to elastically scatter an incoming neutron through any given solid angle in a given time. In effect, when viewing the probabilistic scattering event over all angles, each nucleus in the sample effectively acts as a point source over time, with the scattered neutron propagating and mapping to a spherical wave front (or 'S-wave scattering'). The actual probability of a nucleus to scatter a neutron through any given solid angle per unit time is referred to as a differential cross section, that can be conceptualized as a circle, where the radius relates to what is called the scattering length (in cm) of that specific isotope [35]. The larger the circle, the longer the scattering length and a corresponding

 https://doi.org/10.1051/epjconf/202023603001

increased probability of an elastic neutron-nucleus scattering event (Figure 5). A cross section is essentially a measure of the probability of a nucleus-neutron interaction; there are cross sections that describe the elastic nuclear-potential scattering component, or coherent cross-section – the primary focus of this chapter – as well as elastic spin-scattering (incoherent cross section) and absorption components (indeed, any type of probabilistic interaction of a neutron with a nucleus may be described by a cross section). In the case of a standard SANS experiment, the isotope-specific scattering length composition of a sample per unit volume, i.e., the scattering length density (Figure 5), and the resulting summed convolution of all elastically-scattered neutron amplitudes emanating from each nucleus within the sample is what generates scattering intensities. If the sample has components within it that have preserved point-to-point distance correlations between atomic nuclei internal to their structure, e.g., due to the macromolecule tumbling in solution, the subsequent development of summed elastic-coherent S-wave amplitudes caused by the spatial correlation between all time-preserved scattering centres within the macromolecule is what eventually yields structural information.

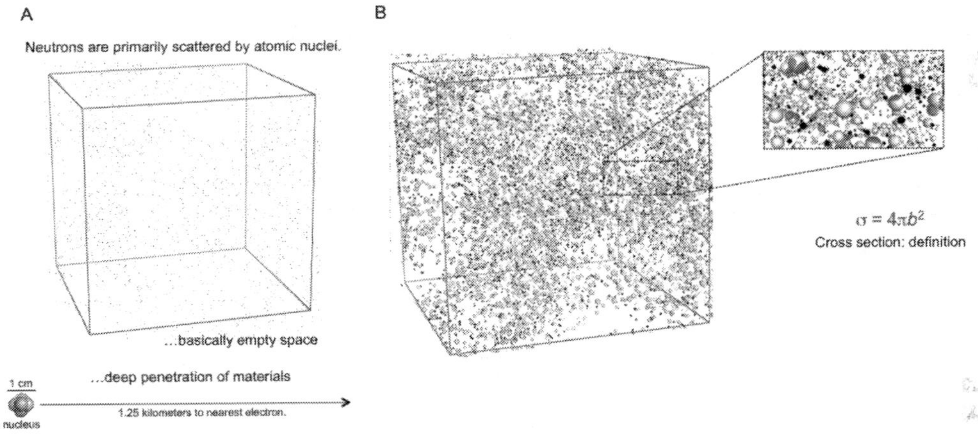

Fig. 5. *What neutrons 'experience' on approaching a block of regular matter.* **A.** Neutrons primarily interact with the nuclei of a material that on an atomic scale are extremely far apart from each other. For example, if a single atom nucleus were scaled up to be 1 cm across, the nearest electron of the same atom would be 1.25 km away. Matter is basically empty space. This is the reason why neutrons are deeply penetrating. **B.** However, each atomic isotope has a different probability to interact with a neutron that can be viewed as 'conceptual circles' occupying the block of matter. These circles are more formally described by the cross section of an isotope, which simply quantifies an atomic nucleus's capacity to interact with a neutron. There are cross sections that describe elastic and inelastic scattering interactions as well as absorption. For neutron scattering, the cross section relates to the square of the scattering length, b (cm), where b is, in effect, the scaled radius of the 'conceptual circles'. Different isotopes have different scattering lengths and therefore different probabilities to interact with an incoming neutron.

One key advantage of SANS is that it is possible to experimentally control the magnitude of the SANS interference pattern. Somewhat intuitively, the SANS intensities can be changed by simply changing the isotopic composition of the sample, i.e., the isotopic composition of the sample-solvent or the macromolecules themselves. Swapping out one isotope for another should, and does, alter both the coherent-scattering S-wave front as well as incoherent spin-scattering contributions. Coincidentally, deuterium (^2H) and the nuclei of commonly occurring isotopes found in biological macromolecules (^{12}C, ^{16}O, ^{14}N, ^{31}P, and mainly ^{32}S) interact with the nuclear potential component to coherently scatter neutrons with a similar magnitude, whereas neutrons scattered from ^{1}H undergo the 'inverse' of what happens compared to the nuclei of ^{2}H and the other biological-elements.

Many of the isotopes encountered in biology have what is termed a positive coherent neutron scattering length, while simple ^{1}H has a *negative* scattering length. A negative scattering length is a terminology that means ^{1}H effectively experiences an attractive energy potential when interacting with neutrons so that the out-going scattered neutron maintains the same phase as it was prior to scattering, i.e., the amplitude of the scattered neutron does not undergo a phase shift relative to the incident neutron (Figure 6). For the nuclei of other biological isotopes and deuterium, a repulsive energy potential interaction with neutrons is set up such that the out-going scattered neutron amplitude undergoes a 180° phase shift (or phase inversion) relative to an incident neutron, i.e., have a positive scattering length. So, neutrons experiencing nuclear-potential scattering from ^{1}H are 180° out of phase with neutrons scattered from most other biological isotopes. Importantly, ^{2}H also has a positive scattering length and therefore the isotopic substitution of ^{1}H for ^{2}H, which alters the average ^{1}H per unit volume of a sample, will systematically alter the overall magnitude of the summed scattering amplitudes: the negative scattering length of ^{1}H is effectively cancelled by the positive scattering length of ^{2}H. SANS can be conceptualized as a 'sum of the waves-game' where the scattering amplitudes can be experimentally controlled by ^{1}H-^{2}H substitution, for example in the supporting background solvent, or buffer (adjusting $^{1}H_2O$:$^{2}H_2O$ ratios) or through the substitution of ^{1}H with ^{2}H at non-exchangeable hydrogen positions within a macromolecule. Note that for exchangeable hydrogens, e.g. those belonging to the hydrophilic groups in the macromolecule, ^{1}H is replaced by ^{2}H proportionally to the percentage of heavy water in solution. Isotopic substitution also affects the magnitude of the incoherent neutron spin-scattering contributions, but in this instance all of the biological isotopes and ^{2}H have, either a positive scattering length, or lack the capacity to undergo spin scattering due to the spin-state of the particular isotope. For biological SANS performed on dilute macromolecular samples in aqueous solution the isotopic substitution of ^{1}H with ^{2}H will result in a decrease in overall spin-scattering contributions as the incoherent scattering length of ^{1}H is enormous in comparison to ^{2}H.

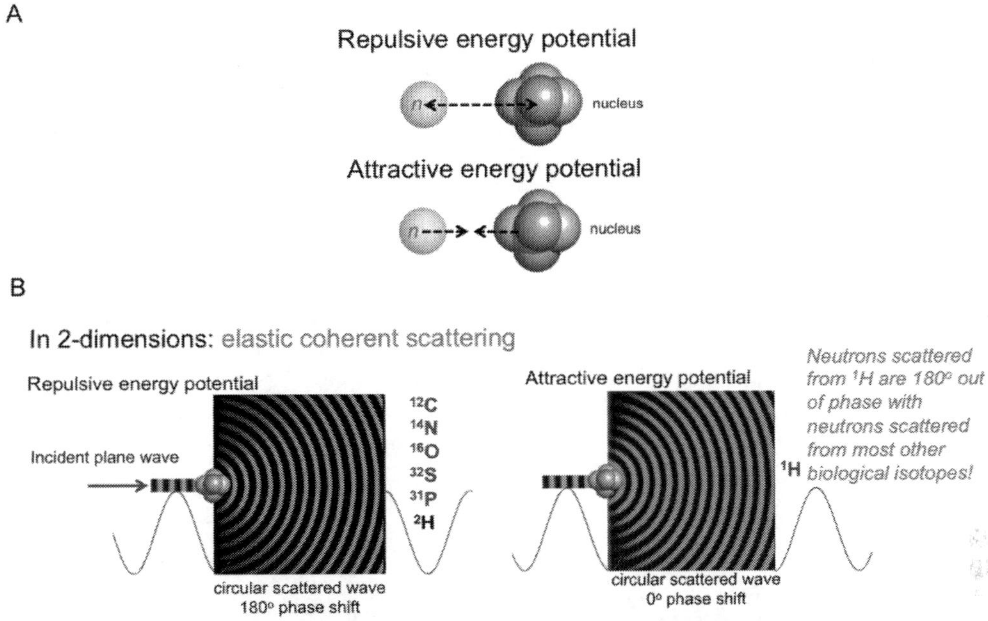

Fig.6. *Nuclear potential interactions and the development of elastic-coherent neutron scattering amplitudes*. **A.** A neutron may experience either repulsive or negative energy potentials when interacting with a specific type of atomic nucleus that is dependent on the isotope. **B.** *Left*: For most of the biological isotopes and deuterium, ^{2}H, neutrons generate a repulsive-potential with the nucleus that results in a 180° 'flip' in the phase of the elastically scattered circular-wave, or S-wave, neutron amplitude relative to the amplitude of the incident neutron. This effect is termed a positive scattering length. *Right*: Neutrons scattered from regular light hydrogen, ^{1}H, experience a repulsive-potential such that the phase-amplitude of the incident neutron is maintained, i.e., the scattered neutron does not undergo an amplitude 'phase-flip' like the other biological isotopes. Therefore, ^{1}H has what is termed a *negative* coherent scattering length, i.e., the scattering amplitudes from ^{1}H are 180° out of phase relative to the neutrons scatted from most of the other biological isotopes.

2.2 The sum of the waves game

The constructive and destructive interference of the scattered neutron amplitudes from each nucleus of a sample gives rise to a convoluted wave front. For a single nucleus, the scattering amplitude remains constant over all angles, only changing in magnitude depending on the scattering length. In other words, the coherent-scattering neutron atomic form factors of the isotopes are effectively angle-independent. However, if *two* scattering centres, related by a time-preserved distance *r*, are present (Figure 7), then lines of constructive and destructive interference are set up in the combined coherent wave front caused by the amplitudes of the two S-waves from each point source interacting and summing together. The most intense line of constructive interference occurs at zero angle, while as the angle increases, the amplitudes undergo angular-dependent 'highs and lows' as the amplitudes arising from both scattering centres constructively and destructively interfere, weighted by the magnitude of the scattering length. The convoluted zero angle amplitudes relate to the combined sum of the individual scattering amplitudes from each scattering centre weighted by the magnitude of the scattering length, while the 'highs and lows' in the amplitudes of the scattered neutron waves at ever-increasing angle relate to the distance separation between the centres, weighted by the scattering lengths. Of course, macromolecules in solution that are tumbling about have many scattering pair distance correlations within the extent of their volume boundary and therefore the coherent wave front from them is derived from the combined sum of the scattering lengths and the scattered wave amplitudes arising from these correlations – time and rotationally averaged – as a function of angle. The lower the angle (lower s), the longer the correlated distances d, such that s and d share a reciprocal relationship, $s = 2\pi/d$, while the summed zero-angle amplitudes reflects the total summed scattering from all scattering pairs within the macromolecular volume, and therefore relates to the molecular mass. It must also be remembered that the solution, i.e., the supporting solvent of the sample, also scatters. As the solvent (hopefully) does not have any time-preserved long-range distances correlations at similar length scales as the macromolecule (e.g., greater than the solvent hydrogen bonding network of 1.5–3 Å), its scattering contributions add as a spatially uncorrelated set of amplitudes that generates a 'flat background' which has to be subtracted to yield the coherent scattering of the macromolecule in solution. Therefore, two measurements have to be performed; on the sample and a corresponding exactly matched buffer to remove the scattering contributions made by the supporting solvent [17].

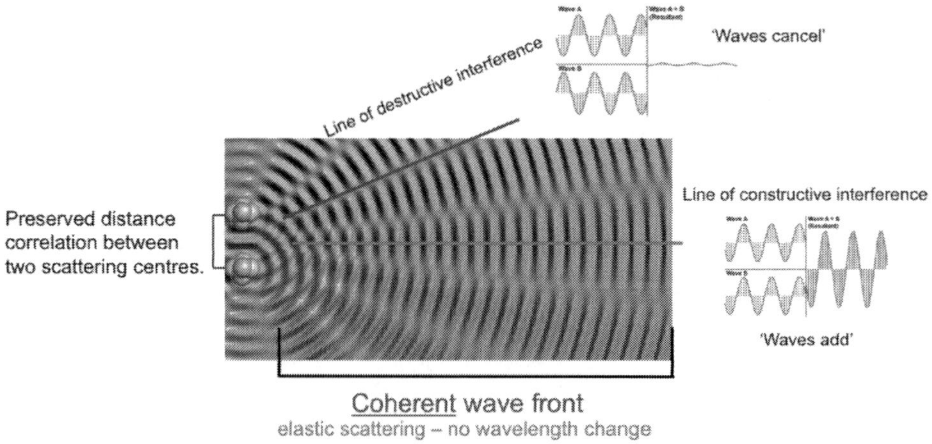

Fig. 7. *The sum of the waves game: preserved distance correlations and amplitude convolution.* If two scattering centres are spatially correlated through time by a distance d, the resulting S-wave scattering amplitudes emanating from each point interfere with each other. The scattering amplitudes, weighted by the scattering length of the isotope at each point, simply add together. The result is a combined coherent spherical wave front of scattering amplitudes that undergo constructive or destructive interference as a function of angle. The 'peaks and troughs' in the summed amplitude wave relate to the initial distance between the scattering centres. Unfortunately, it is not possible to access these scattering amplitudes experimentally: SANS records the intensity of the scattered neutrons that manifest as the square of the scattering amplitudes.

2.3 From nuclei to macromolecules: From reciprocal space amplitudes to real-space distances.

More formally, an incident neutron that makes up a collimated beam that travels toward a sample in a direction z, has what is termed a wave vector, $\boldsymbol{k}$, that also manifests in the same direction and relates to the neutron energy, i.e., its wavelength. Consequently, the incident neutron that is in effect collimated as a plane wave can be described by a complex-valued wave function:

$$\varphi_{incident} = e^{ikz} \quad (1).$$

When a nucleus gets in the way of this incident neutron and an elastic scattering event occurs, (i.e., the magnitude of the scattering vector, $\boldsymbol{k}$, does not change), then the resulting scattered neutron also follows a wave function weighted by the probability of a nucleus to scatter the incident neutron in any direction r, through a given time that when viewed over all scattering angles can be conceptualized as a 'probabilistic spherical/circular wave' (or the 'e^{ikr}' component in the following expression):

$$\varphi_{scattered} = -\frac{b}{r}e^{ikr} \quad (2).$$

The scattered neutron wavefunction is nominally negative to reflect the fact that most isotopes cause a phase inversion in the amplitudes of the incident neutron, i.e., have a positive scattering length, b, arising from the repulsive potential interaction between the incident neutron and nucleus. In turn, the scattering length of an isotope relates to the total scattering cross section, σ_c, which is a measure of the total number of neutrons scattered per second by that nucleus, normalized to the incident neutron flux, or perhaps more simply: The effective probability of a particular nucleus to scatter a neutron in any given time in any direction is:

$$\sigma_c = 4\pi b^2 \quad (3).$$

The magnitude of the scattering length, which is isotope specific, will then reflect the neutron 'scattering ability' for each nucleus and the production of the resulting spherical/circular-scattered wave amplitudes. Intuitively, if there are multiple isotopes in a sample, then the sum of their different scattering abilities, and the corresponding wavefunctions, will add together – for example, substituting ^{1}H with ^{2}H, i.e., from a negative b to a positive b, will radically alter the combined $\varphi_{scattered}$ i.e., the magnitude of the amplitudes of the scattered wave front.

As mentioned above, depending on the isotope in question, their initial energy states, the energy of the incident neutrons, energy transfer and spin interactions that occur within the compound neutron-nucleus, nuclear resonance states develop that lead to:

- Scattering (the real component of b).
- Absorption (imaginary components of b).

For the purposes of this chapter, absorption will not be described in detail, but basically involves the interaction of a neutron with a particular incident energy and a nucleus leading to the formation of a compound-neutron-nucleus that is commensurate in energy with an excited nuclear resonance-state (i.e., the compound-nucleus has a large absorption cross section). For most applications in biological SANS, the imaginary absorption component of the scattering length is small, i.e., the interactions within the compound-nucleus are away from the excited resonance states. The scattering events therefore arise from the real component of b. The total neutron scattering length, in turn, consists of two terms:

$$b_n = b_p + b_s \quad (4),$$

where the b_p represents the probabilistic interaction of a neutron with a nuclear potential – the coherent scattering length – and b_s represents its interaction with the nuclear spin (Figure 8).

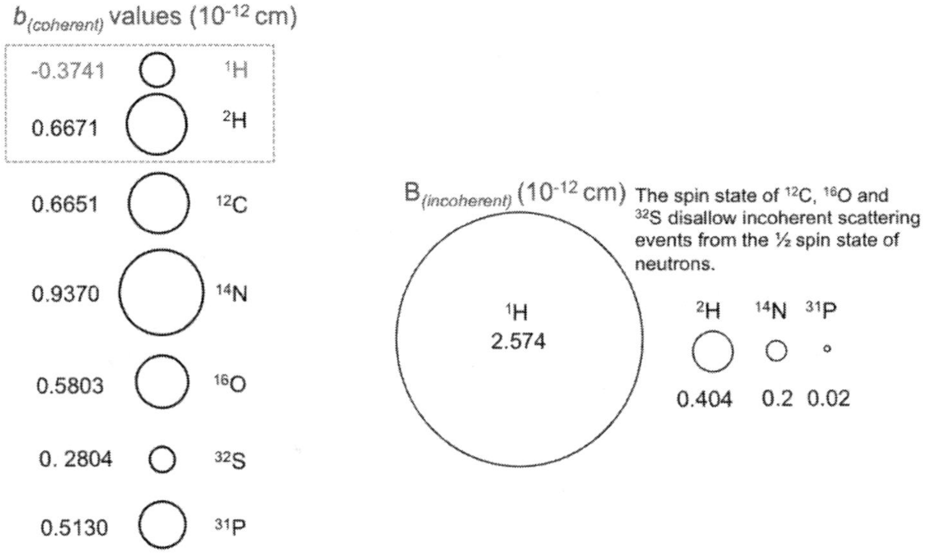

Fig. 8. *Neutron scattering lengths of common biological isotopes*. The magnitude of neutron scattering cross-sections represented as circles.

For SANS, the neutron-nuclear potential interaction, b_p, is important as this gives rise to elastic coherent scattering events that can be used to determine the atom-pair separation within a macromolecule. The spin scattering term, b_s, adds to this signal and can be expressed as:

$$b_s = 2\mathrm{B}(\boldsymbol{IS}) \quad (5),$$

where B is the spin scattering length of an isotope ($\boldsymbol{I}$ and S are spin operators of the neutron and nucleus). This additional spin-interaction and the resulting elastic scattering produced by it can cause a problem for biological SANS experiments. If the spins of the atoms comprising the sample and of the incoming neutrons are not ordered, neutrons will scatter incoherently. In other words, correlations between the wave-amplitudes of the spin-scattered neutrons and the distances between the spin-scattering centers no longer exist. Incoherent elastic spin-scattering produces additional, and sometimes significant, background scattering contributions depending on the isotope, and in particular light-hydrogen. In simple terms, the incoherent scattering amplitudes 'ripple across' the entire coherent amplitude wave front, convoluting with them across all angles to generate statistical fluctuations in the coherent amplitudes that manifest as noise in the measured SANS intensities. As most biological samples are weakly-scattering, are dominated by ^{1}H and are typically measured in a dilute-aqueous regime, without a spin-polarized neutron source or oriented nuclear spins, incoherent scattering 'noise' can come to dominate a scattering profile necessitating long exposure times to improve signal-to-noise ratios. As can be seen in Figure 8, regular ^{1}H is unusual compared to the other biological-isotopes in that its coherent scattering length is negative, and its incoherent scattering length is enormous. The practical consequences are that solvent-subtracted SANS patterns measured from macromolecules in solution in buffers containing high percentages of ^{1}H are, when measured for the same amount of time, quite noisy compared to samples in high percentages of ^2H_2O, i.e., where ^1H has been swapped out for ^2H in the buffer.

If the distance vectors, $\vec{r}$, between any two atoms are spatially correlated, then the magnitude of the amplitudes, A, of the convoluted S-wave amplitudes as a function of s (Figure 7) are proportionate to the sum of the individual atomic scattering factors for each isotopic nucleus (i.e., probability to scatter):

$$A(s) = \sum_{i=1}^{N} b_i e^{i\vec{s}\vec{r}} \qquad (6).$$

The issue is that it is not possible to access the scattering amplitudes experimentally. SANS measures the intensity of the scattered neutrons, that is, the scattering amplitudes multiplied by their complex conjugate, or more simply, the absolute magnitude of the wave amplitudes squared:

$$I(s) = A(s)A(s)^* = |A(s)|^2 \quad (7).$$

Fundamentally, at zero angle, $I(0)$, the amplitudes from both nuclei will have a maximum-level of 'in-phase constructive interference'. For macromolecules that contain numerous spatially-correlated scattering pairs – and after background scattering contributions have been subtracted – the combined sum of the squared amplitudes that manifest experimentally as $I(s)$ vs. s, will depend on the number and type of nuclei per unit volume, i.e., the scattering length density, ρ, as well as the real-space distribution of distances, $\vec{r}$, between all scattering pairs. Different isotopes will have different scattering lengths resulting in different weighted contributions to $I(s)$, while the real-space distance-separation frequency within the volume boundary of a macromolecule will determine the angular dependence in the intensities, i.e., the constructive and destructive interference pattern of the squared scattering amplitudes as a function of angle. For a continuous *excess* neutron scattering length density distribution, $\rho(\vec{r})$, bounded by the total particle volume, V_r, the resulting form factor, $F(s)$, of a macromolecule describes the magnitude and subsequent convolution of the scattering amplitudes (eq.6) caused by time-preserved real-space nuclei-pair separations (eq.7) and can be expressed as:

$$I(s) = \langle F(s)F(s)^* \rangle_\Omega = \langle |F(s)|^2 \rangle_\Omega, \qquad \text{where } F(s) = \langle \int_{V_r} \rho(\vec{r}) e^{i\vec{s}\vec{r}} dV_r \rangle_\Omega \quad (8),$$

The $\langle \quad \rangle_\Omega$ in the above expression represents that for a pure, infinitely-dilute macromolecule tumbling in solution (Ω), the resulting form factor is rotationally averaged over all orientations, yielding an isotropic scattering pattern. Eq.8 shows that the $F(s)$ of a macromolecule and the distribution of the excess scattering length density $\rho(\vec{r})$ are related by a Fourier transform. It is possible to express the distribution of $\rho(\vec{r})$ in terms of an autocorrelation function, $\gamma(r)$, that is essentially the spatial convolution of the excess scattering length density of an object with itself taking all pair-wise distances within the volume boundary into account over all orientations calculated from a common origin (Figure 9). The $I(s)$ relates to this autocorrelation function via:

$$I(s) = \langle \int_{V_r} \gamma(r) e^{i\vec{s}\vec{r}} dV_r \rangle_\Omega \quad (9).$$

The coherent S-wave amplitude components of the relation, $\langle e^{isr}\rangle$, can be re-formulated as:

$$\langle e^{i\vec{s}\vec{r}}\rangle = \frac{\sin(sr)}{sr} \quad (10),$$

or an 'integrated sin(x)/x Fourier term', while the integral of all scattering-pair volume elements, dV_r, can be expanded and then simplified to;

$$dV_r = 4\pi r^2 dr \quad (11).$$

The substitution of eq. 10 and 11 into eq. 9 produces the relation between I(s) and $\gamma(r)$:

$$I(s) = \int_0^\infty 4\pi\gamma(r)\frac{\sin(sr)}{sr}r^2 dr \quad (12).$$

What does all of this mean? Fundamentally what these relationships describe that it is possible to obtain the probable frequency of all scattering pair-distances distances, r, within a macromolecule from $I(s)$ vs. s, i.e., to transform between the measured intensities in momentum- or 'reciprocal-space' into real-space (Figure 9). Therefore, the measured coherent SANS intensities should reflect the time-preserved excess scattering length density distribution within a single macromolecule, i.e., the structure. The relationship between $I(s)$ and the r distance correlations can then be expressed as a probable real-space scattering-pair distance distribution, or $p(r)$. As:

$$p(r) = \gamma(r)r^2 \quad (13),$$

Then substitution into eq. 12 yields $I(s)$ in terms of $p(r)$:

$$I(s) = 4\pi\int_0^\infty p(r)\frac{\sin(sr)}{sr}dr \quad (14),$$

Which on inverse transformation leads to the expression of the real-space distance distribution in terms of $I(s)$:

$$p(r) = \frac{r^2}{2\pi^2}\int_0^\infty s^2 I(s)\frac{\sin(sr)}{sr}ds \quad (15).$$

The scattering intensities – and thus the associated form factor amplitudes of the entire macromolecule in reciprocal space – relate to the frequency of atom-pair distances in the macromolecule in real-space by a Fourier transform.

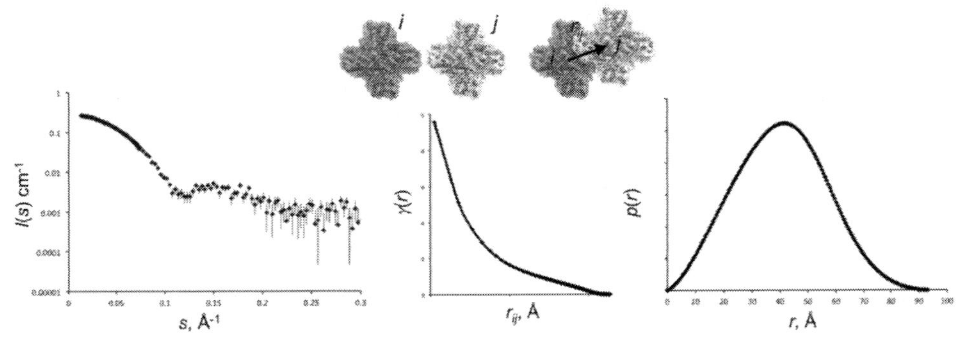

$$I(s) = 4\pi \int_0^{D_{max}} p(r) \frac{\sin(sr)}{sr} dr \qquad p(r) = \frac{r^2}{2\pi^2} \int_0^{\infty} s^2 I(s) \frac{\sin(sr)}{sr} ds$$

Fig. 9. *Scattering intensities in relation to real-space scattering-pair frequencies.* If a macromolecule is pure, monodisperse, and is sufficiently dilute so as does not experience any interparticle interactions, the resulting background-corrected scattering profile recorded in reciprocal space, $I(s)$ vs. s, relates to the probable frequency of preserved scattering-pair distances, r, in real space, $p(r)$, via a Fourier transform. In effect, the resulting $p(r)$ represents the overall structure of the macromolecule expressed in terms of a distance distribution and can be derived from the excess scattering length density distribution internal to the particle, here represented as $\gamma(r)$. The $\gamma(r)$ can be viewed as the spatial correlation of the particle, i, with itself, j, in terms of the autocorrelation of all r_{ij} vector-pairs determined from a set origin and over all particle orientations, the maximum extent of which is determined by the maximum particle dimension ($\gamma(D_{max}) = 0$).

It is of note here that the conversion between reciprocal-space scattering and a real-space distance distribution is not a trivial exercise using the relationship above (eq.15) due to the discrete properties of the experimental data that are otherwise measured in Δs increments within a limited experimental s-range (s_{min}–s_{max}), and the fact that macromolecules have a maximum particle dimension, D_{max}, that provides an upper limit of r. Through these restrictions into the 'sin(x)/x Fourier term' as well as experimental noise, i.e., the scattering variance, the resulting solution(s) of $p(r)$ can become unstable. The nature of the SANS measurement does not satisfy continuous integral/mathematical expressions. Therefore, it is necessary to employ indirect inverse Fourier transform methods to obtain $p(r)$, for example implemented in the programs GNOM[27] or GIFT[28]. Here, $p(r)$ is described in terms of a linear combination of coefficient-weighted functions, $\phi(r)$ across the interval of $r = 0$–D_{max} whose individual Fourier transforms $\psi(s)$ are known. The experimental data and the resulting $p(r)$ are represented as a coefficient-weighted sum of these functions:

$$p(r) = \sum_{k=1}^{K} c_k \varphi_k(r) \qquad (16).$$

$$I_{calc}(s) = \sum_{k=1}^{K} c_k \psi_k(s) \qquad (16a).$$

The derived $p(r)$ profile is optimized by the coefficients, c_k, that multiply each $\phi(r)$ so as to minimize the reduced χ^2 discrepancy between the experimental data points, $I_{exp}(s_j)$, and the calculated scattering intensities, $I_{calc}(s_j)$, taking into account the experimental errors, $\sigma_{exp}(s_j)$, while also maintaining a smooth $p(r)$ in real-space by incorporating a penalty term $P(p)$. The function to be minimized is:

$$\Psi = \chi^2 + \alpha P(p) \quad (17),$$

where:

$$\chi^2 = \frac{1}{N-1}\sum_{j=1}^{N}\left[\frac{I_{\text{exp}}(s_j) - cI_{\text{calc}}(s_j)}{\sigma(s_j)}\right]^2 \quad (18),$$

(c is a scaling constant, N is the number of points), and:

$$P(p) = \int_{0}^{D_{max}} \left[\frac{\mathrm{d}p}{\mathrm{d}r}\right]^2 \mathrm{d}r \quad (19).$$

Relationship (19) describes a smoothing term applied to $p(r)$ in real-space where the regularization parameter α in eq. 17 acts as a balance between the stability of the computed $p(r)$ and the goodness of fit to the experimental data in reciprocal-space. For example, improving χ^2, so as to minimize eq.18, would mean the calculated fit to the data starts to trace every experimental point, i.e., the resulting $p(r)$ in real-space would start to become increasingly affected by the sampling variance in the experimental intensities, resulting in an unphysical $p(r)$ profile with many maxima and minima (i.e., it would be overly-affected by experimental noise). In this case α is too low. Conversely, attempting to limit the influence of experimental noise by increasing α, i.e., increasing the magnitude of the smoothing term, may generate a smooth $p(r)$ but a poor fit to the data. The programs GNOM[27] or GIFT[28] help in the selection of an appropriate stable solution of $p(r)$ in real-space and the corresponding optimized fit to the data in reciprocal space. These relationships and the subsequent 'discrete/indirect approach' to what is otherwise an ill-posed problem of the inverse-Fourier transform described by the continuous integrals of eq.14 and 15, simply show that it is possible to extract real-space scattering-pair distance information within the constraints imposed by the experimental data.

2.4 Solutions, macromolecules and contrast.

The resultant sum of the coherent neutron scattering amplitudes and subsequent scattering intensities produced from the macromolecules making up a sample are generated because of the spatial correlation between scattering pairs within each macromolecule. In the small-angle regime, it is often the case that the buffer/solvent employed – and unlike the macromolecules themselves – lack these time-preserved long-range distance correlations. Consequently, there is no effective pair-wise 'self-amplification' of the scattering amplitudes arising from the solvent and, as a result, the solvent scattering scale in intensity across all angles only based on the average scattering length density, i.e., the isotopic composition per unit volume and the summed 'scattering power' of each isotope (eq.4). This results in a 'flat' scattering contribution made by the solvent across a SANS profile that can be subtracted from the sample scattering by measuring an equivalent solvent blank (Figure 3). The background-corrected *I*(*s*) vs. *s* profile will then represent the form factor of the population of macromolecules in solution.

Importantly, *I*(*s*) in the small-angle region only arises if there is an excess scattering length density within the particle volume relative to the solvent, that is, there is a difference between the average scattering length density of the solvent, $\bar{\rho}_s$, and the average scattering length density of the particles of interest, $\bar{\rho}$. This difference is called contrast ($\Delta\rho$: Figure 10):

$$\Delta\rho = V_r^{-1}\int_{V_r} (\bar{\rho}(\vec{r}) - \bar{\rho}_s)dV_r = \bar{\rho} - \bar{\rho}_s \quad (20).$$

It is reasonably straightforward to see that if the scattering amplitudes arising from the solvent are of a similar magnitude as the macromolecule floating around in it, the resulting scattering intensities will be 'drowned out' by the solvent scattering, i.e., where $\Delta\rho$ limits to zero. However, for SANS, the ability to alter the contrast is the key to obtain structural information between components of complexes and assemblies of *different* average scattering length density because it is possible to match-out the scattering contributions of selected components by altering the scattering length density of the solvent.

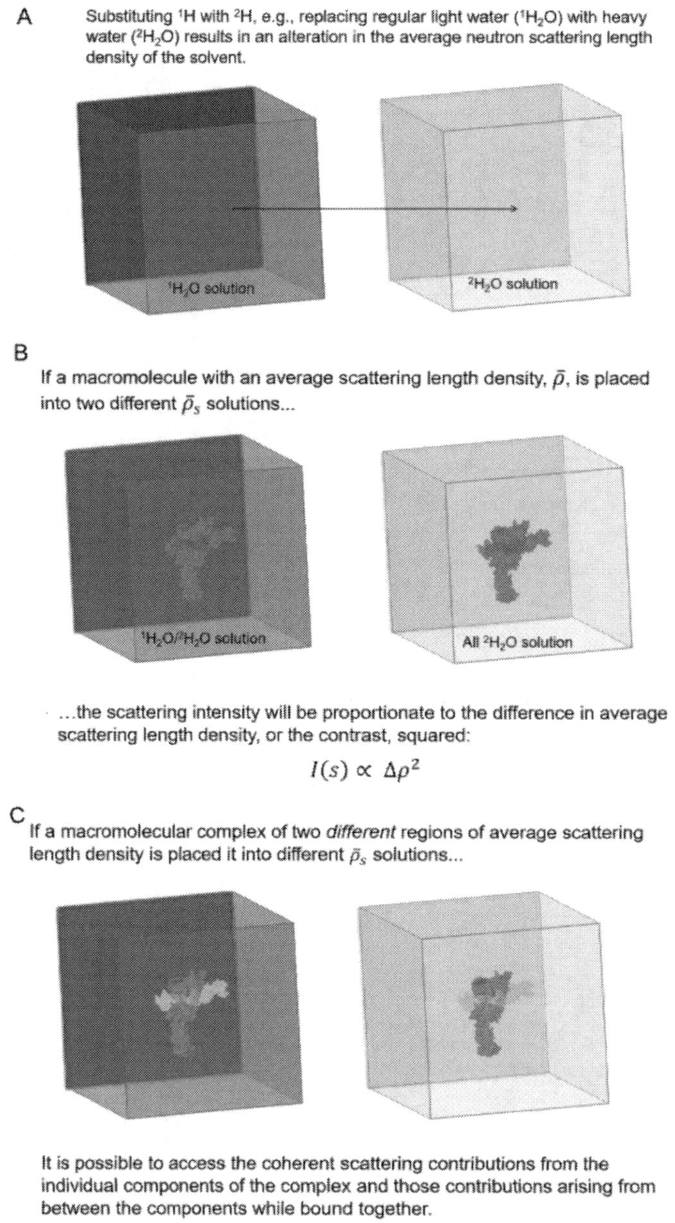

Fig. 10. *Solutions macromolecules and contrast.* **A.** Altering the isotopic composition of a solvent changes the average neutron scattering length density. For example, as the scattering length of ^{1}H is negative, swapping ^{1}H with ^{2}H in the solvent, that has a positive scattering length, changes the scattering length density considerably. **B.** Placing a macromolecule into a solvent with a similar average neutron scattering length density as the solvent results in poor contrast, $\Delta\rho$. Changing the scattering length density of the solvent or macromolecule via ^{1}H-^{2}H substitution radically alters the contrast and the resulting magnitude of the scattering intensities as $I(s)$ a $\Delta\rho^2$. **C.** For macromolecular complexes with two regions of average scattering length density, altering the contrast of the supporting solvent enables components to be ‘matched-in’ and ‘matched out’ of a SANS profile yielding structural information from the complex, the individual components and scattering contributions from between the components (the cross-term).

If we take a solution, we can sum the coherent scattering lengths for each isotope of each atom per unit volume to obtain the mean scattering length density [36]. Intuitively, as the scattering amplitudes of ^1H are 180° out of phase with ^2H, then substituting ^1H with ^2H in the solvent e.g., replacing regular light water (^1H_2O) with heavy water (^2H_2O,) will alter the average scattering density of that solvent; the positive scattering length of ^2H and the other isotopes in the solution will cancel-out the negative scattering length of ^1H as the concentration of ^2H_2O is increased (Figure 10). If we take a macromolecule, that typically has reasonably stable level of non-exchangeable ^1H per unit volume (i.e., ^1H covalently bound to functional groups that are not in short-time scale exchange with the solvent) and place it into a solution with a different average scattering length density, the magnitude of scattering intensities will be proportional to the contrast squared, i.e.,

$$I(s) \propto \Delta\rho^2 \quad (21).$$

Consequently, when systematically swapping out ^1H_2O with ^2H_2O there will come a point where the average ^1H per unit volume in the solvent is such that the resulting scattering length density will, on average, be equivalent to the average scattering length density of the macromolecule (i.e., the scattering amplitudes arising from non-exchangeable ^1H *plus* all of the other biological isotopes making up the macromolecule will be cancelled by the scattering amplitudes from the solvent). As a result, the contrast will equal zero, eliminating the scattering signal at zero angle (note that at non-zero angles there will be non-zero scattering due to particle structure, see Section 4). The concept of contrast matching applies irrespective of the size of the macromolecule under investigation. If an enormous macromolecule (with a large volume) is placed in a $^1H_2O/^2H_2O$ solution such that $\Delta\rho = 0$, the forward scattering intensities will limit to zero, as it applies to small macromolecules in solution.

Contrast matching becomes a very useful property for any SANS investigation. Different classes of macromolecule have different average isotopic compositions per unit volume i.e., have different average scattering length densities due to their different chemical compositions. For example, protein, DNA, carbohydrates and lipids are comprised of different ^1H:^{12}C:^{16}O:^{14}N:^{31}P: and ^{32}S ratios. Consequently, proteins typically have coherent neutron scattering match points between 40–43% v/v ^2H_2O, whereas lipids match out between 2–15% v/v ^2H_2O and DNA matches out at ~60-70% v/v ^2H_2O (Figure 11). Therefore, if hetero-macromolecular complex with two regions of average scattering length density is placed into the appropriate % v/v $^1H_2O/^2H_2O$ solvents, it is possible to extract the scattering contributions for the whole complex (e.g., 100% v/v ^1H_2O buffers) and from the individual components. For example, a protein/DNA complex will have three contrast match points of around 43% v/v ^2H_2O for the protein component (whereby the scattering signal is dominated by contributions from the DNA) and around 65% v/v ^2H_2O for the DNA component (where the scattering signal will be dominated by the protein) as well as the a whole-complex match point (whereby the entire complex is matched out from the scattering signal). Therefore, using either contrast matching, or contrast variation, it should be possible to extract structural information from complexes with different regions of average scattering length density and obtain distance distributions of the individual components as well as the distances between those components of the complex (called the 'cross-term'). SANS is one of the few techniques that enables the determination of the orientation and relative disposition of components within a full-formed macromolecular complex in solution and to assess whether, for example, any major structural alterations occur in one or more the components on forming a complex.

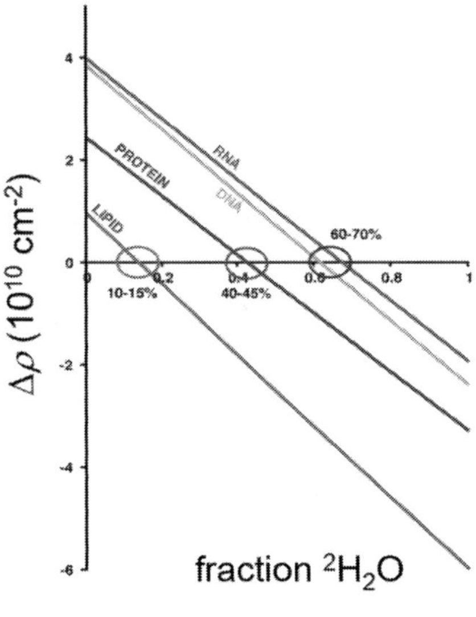

Fig. 11. *Biomacromolecule classes and their average neutron scattering contrasts in ^2H_2O.* Proteins, lipids, and polynucleotides have different average neutron scattering length density due to different chemical/isotopic compositions. These contrast in volume fractions of ^2H_2O can be altered by isotopically labelling biomolecules with different levels of non-exchangeable ^2H, e.g., using biodeuteration. This will shift the match-point of the macromolecule, i.e., where the contrast limits to zero.

The capacity to alter the contrast in a SANS experiment via ^1H-^2H substitution opens up a vast array of experiments. Not only is it possible to alter the scattering length density of the solvent, but it is also possible to control the average scattering length density of macromolecules using, for example, recombinant protein expression combined with non-exchangeable ^2H isotopic labelling [17]. Such experiments allow for the interrogation of unlabelled/labelled protein-protein complexes and their assemblies. In addition, lipids or detergents that are used in membrane protein investigations, can be chemically synthesized with different scattering length densities to effectively render them invisible in a contrast-matching/variation SANS experiments, affording a means through which to determine the low-resolution structure of a proteins and complexes embedded in detergent micelles or multi-component lipid nanodiscs [37, 38] or bicelles [55-56]. Nanoparticle conjugates can also be investigated, e.g., the covalent attachment of proteins, lipids or organic polymers to heavy-metal nanoparticles [39, 40]. The coherent neutron scattering length of, for example gold (0.760 10^{-12} cm) or iron (0.945 10^{-12} cm) is of a similar magnitude as ^2H (0.667 10^{-12} cm), and therefore these elements can be contrast-matched in high concentrations of ^2H-containing solvents, leaving behind SANS signal derived from any ^1H-organic conjugate that will otherwise have high-contrast due to the negative scattering length of ^1H (-0.374 10^{-12} cm).

2.5 Solution SANS in summary.

To obtain structural information from macromolecules SANS, the solvent scattering contributions must be subtracted to yield the excess scattering length density of the macromolecule in solution, due to a difference in contrast. As would be expected, the larger the volume of a macromolecule, the greater number of scattering-pair distances will be present within the volume boundary, giving rise to a more intense contrast-weighted scattering profile at low angles. This leads to a generalized expression that relates the magnitude of the measured scattering intensities to the bulk properties of a sample with the form-factors of each individual macromolecule, i, of a population:

$$I(s) = S(s) \sum_{i}^{n} \left[(\Delta\rho_i V_i)^2 \langle |F(s)|^2 \rangle_{\Omega_i} \right] \quad (22).$$

The measured scattering intensities simply the summed contribution of all individual macromolecule form factors, weighted by the product of the volume squared and contrast squared relative to the solvent. The first term, called the structure factor, $S(s)$, relates to the contributions to $I(s)$ made by correlated distances of closest approach between the macromolecules in the sample. As long as a solution SANS experiment is performed using a sufficiently dilute sample, the $S(s)$ (also called interparticle interference) limits to 1. Therefore, and on the condition that each and every molecule within the population is identical, i.e., is pure, monodisperse and non-interacting, then eq. 22 simplifies to:

$$I(s) = N(\Delta\rho V)^2 \langle |F(s)|^2 \rangle_{\Omega} \quad (23),$$

such that the measured $I(s)$ profile after background correction will represent the time and rotationally averaged scattering from a single particle in solution (be it a single macromolecule, a complex, higher-order assembly, nanoconjugate, etc.). By extension, for a multi-component and monodisperse complex or assembly that contains *two* regions of average scattering length density (e.g. a protein/nucleic acid complex), the contributions made to the buffer-subtracted scattering profile by the two regions follow:

$$I(s) = N[(\Delta\rho_1 V_1)^2 \langle |F_1(s)|^2 \rangle_{\Omega} + (\Delta\rho_2 V_2)^2 \langle |F_2(s)|^2 \rangle_{\Omega} + 2\Delta\rho_1 V_1 \Delta\rho_2 V_2 \langle |F_{12}(s)|^2 \rangle_{\Omega}] \ (24)$$

that is, the total (coherent) scattering intensities incorporate the contrast-weighted scattering from: Component 1 ($\Delta\rho_1 V_1$); component 2 ($\Delta\rho_2 V_2$) and; the component 1-2 scattering cross-term resulting from scattering events between the different scattering-length density regions ($\Delta\rho_1 V_1 \Delta\rho_2 V_2$). In effect, what this relationship shows is that by controlling the scattering length density of the sample by taking advantage of ^1H's 'negative phasing power' either via $^1H_2O/^2H_2O$ substitution in the solvent and/or swapping non-exchangeable ^1H with non-exchangeable ^2H in the macromolecule, will yield structural parameters relating to the form factor of the assembly ($F_1(s) + F_2(s) + F_{12}(s)$), the individual components of the assembly ($F_1(s)$ or $F_2(s)$) and the between component cross-term ($F_{12}(s)$). All of this structural information relies on those conditions where $\Delta\rho = 0$ at the requisite component match points.

3 Structural Parameters.

There are two key structural parameters that can be extracted from SANS data that relate to the contrast times volume of a macromolecule and the scattering length distribution within this volume, expressed as the radius of gyration, R_g. If an experiment is performed on a complex consisting of two components of different average scattering length density and contrast variation is employed, these structural parameters vary systematically as the coherent scattering contributions of the components are 'matched out' ($\Delta\rho = 0$) or 'matched in' ($\Delta\rho \neq 0$) as the $^1H_2O/^2H_2O$ ratio in the supporting solvent is adjusted.

3.1 Forward scattering, I(0).

At zero angle, the magnitude of $I(s)$ primarily depends on the total number of scattering centres within the squared-volume of a macromolecule – independent of the shape – weighted by the number density, N, (or sample concentration) and the contrast squared:

$$I(0) \sim N(\Delta\rho V)^2 \quad (25).$$

The intensity at zero angle, also called forward scattering $I(0)$, is the total contrast-weighted scattering derived from the constructive interference of the coherent scattering amplitudes arising from all scattering pairs within the volume boundary of the particles under the assumption that there are no interparticle interference effects. Of course, $I(0)$ is not accessible experimentally as the zero-angle position is occupied by the incident neutron beam. However, $I(0)$ can be calculated using the Guinier approximation (see 3.2 below) or from the area under $p(r)$, from eq.15:

$$I(0) = 4\pi \int_0^{D_{max}} p(r)dr \quad (26).$$

If the system under study experiences a change in contrast by altering the $^1H_2O/^2H_2O$ ratio of the solvent, then the magnitude of $I(0)$ will also change. The square-root of $I(0)$, when normalized to the sample concentration, should decrease linearly as the fraction of ^2H_2O increases to the point where $\Delta\rho = 0$ and a whole particle scattering contribution is matched out from the SANS profile. As the fraction of ^2H_2O increases even further, the contrast term goes negative, resulting in an increase in $I(0)$ once again (Figure 12). As $I(0)$ is linked to the V^2 of a macromolecule, and as macromolecules have an average scattering length density dependent on the isotopic composition, then it is possible to obtain the molecular weight (MW) of the macromolecule in solution using the following relationship:

$$MW = \frac{I(0)N_A}{c(\Delta\rho v)^2} \quad (27),$$

where N_A is Avogadro's number, υ is the partial specific volume (cm^3/g, [36]) and c the concentration in g/cm^3. Importantly, for this relation to hold, the SANS scattering intensities need to be measured on an absolute scale, cm^{-1}, which is routine practice at SANS beam lines. Deviations in the experimentally determined MW compared to an expected MW of a macromolecule (e.g., calculated from the atomic composition), or deviations from linearity in the $\sqrt{I(0)}$ vs fraction of ^2H_2O leading up to or away from the whole-particle match-point[52,53] are key indicators that the macromolecules in a sample are being perturbed, for example may aggregate in high volume fractions of ^2H_2O, or in the case of multi-component systems, experience component disassociation.

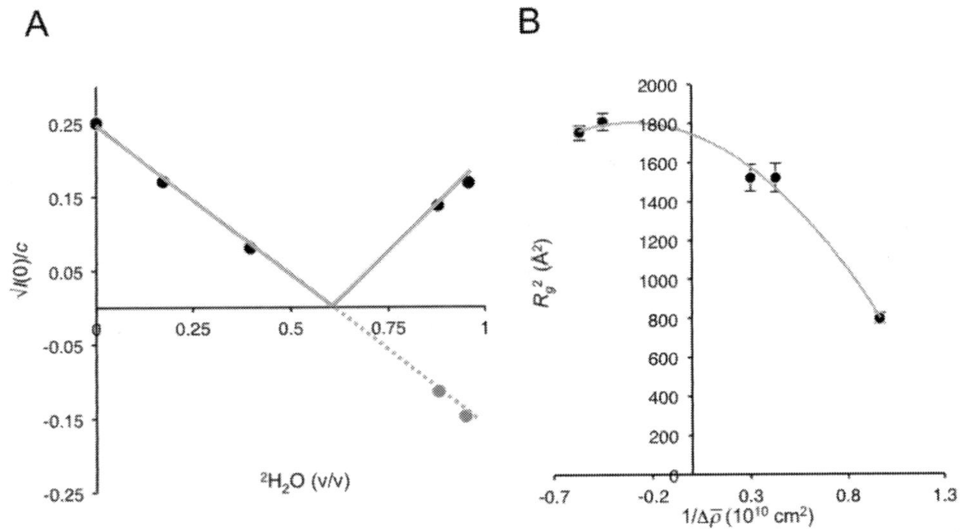

Fig. 12. $I(0)$ *and* R_g *through a contrast variation series.* **A.** The forward scattering $I(0)$, when normalized to sample concentration, c, systematically changes as the average neutron scattering length density of the solvent is altered by increasing the ^2H_2O concentration. As $I(s) \propto \Delta\rho^2$, a plot of $\sqrt{I(0)/c}$ vs. the volume fraction of ^2H_2O in the solvent should generate a straight line leading up to and away from the match point of the entire macromolecule ($\Delta\rho = 0$), in this instance where the volume fraction of ^2H_2O = 0.55. To assess whether high concentrations of ^2H_2O perturb the sample, it is common practice to multiply the experimental $\sqrt{I(0)/c}$ obtained above $\Delta\rho = 0$ by -1 to generate a straight line (dashed-grey) and estimate the linearity of the resulting plot across all contrast points. Deviations from linearity suggest something is happening in the sample, e.g., aggregation or the disassociation of that can be cross-checked against concentration-dependent molecular weight estimates of the sample at each contrast point (eq. 27). **B.** Stuhrmann plot showing the relationship between the inverse of the contrast and the R_g^2 and as described in eq. 30.

3.2 The Guinier approximation, radius of gyration, and Stuhrmann relationship.

The R_g is a measure of the contrast weighted root mean square (or quadratic mean) distance of all volume elements occupied by scattering centres with respect to the centre of the scattering length density of a particle. This centre may *or may not* coincide with the centre of mass of a macromolecule or complex. The magnitude of the R_g is a useful overall parameter influenced by the size, shape and internal structure of the macromolecule. For example, an observed change in R_g of 2 Å for a compact-globular particle equates to, either a significant change in shape, or, alternatively, a significant change in volume corresponding to a sizeable increase in mass.

Guinier [25, 26] showed that the scattering intensity at very-low angles is dependent on the radius of gyration:

$$I(s) = I(0)e^{\frac{-s^2 R_g^2}{3}} \quad (28).$$

For monodisperse macromolecules in solution, and after the solvent scattering contributions have been subtracted, a plot of the natural logarithm of the experimental $I(s)$ vs s^2 – or Guinier plot –will produce a negative linear relationship in the range $s < s_{min}$ where $s_{min}\, R_g < 1.3$. When extrapolated to the ln$I(s)$ intercept at $s = 0$, this yields $I(0)$ and the slope of the plot is proportional to R_g^2. The R_g may also be calculated from the second moment of $p(r)$, eq. 15):

$$R_g^2 = \frac{\int_0^{D_{max}} p(r)r^2 dr}{2\int_0^{D_{max}} p(r)dr} \quad (29).$$

The two independent R_g estimates from the Guinier approximation and $p(r)$ may be compared and cross-checked for consistency. However, the R_g as determined from the Guinier approximation, as opposed to $p(r)$, may become increasingly difficult to assess for structurally anisotropic inhomogeneous systems consisting of distinct regions of neutron scattering length density under conditions of varying contrast, i.e., when a complex has internal scattering-length density heterogeneity.

The proposition arises that for macromolecular complexes with two regions of average scattering length density, the R_g must be affected by $\Delta\rho$ and it should be possible to evaluate the R_g of a whole complex and the R_g of the individual components of the complex as the $\Delta\rho$ is altered. In addition, it should also be possible to assess the systematic change in R_g as a function of contrast that ultimately relates to the disposition between the two region centres of mass internal to the complex as the component scattering contributions are systematically 'phased-in' and 'phased-out' of a SANS profile as a function of % v/v ^2H_2O in the solvent.

Stuhrmann [52,53] showed that a relationship exists between the R_g^2 and the inverse of the contrast for a scattering object with an internal heterogeneous scattering length density (Figure 12):

$$R_g^2 = R_m^2 + \frac{\alpha}{\Delta\rho} - \frac{\beta}{\Delta\rho^2} \quad (30),$$

where R_m is the radius of gyration of the particle at 'an infinite contrast'. At infinite contrast, the last two terms of the Stuhrmann relationship limit to zero. Hypothetically, at this point, the difference in scattering length density between heterogeneous regions would, in principle, also limit to zero (i.e., for an n-component particle, $\Delta\rho_n = \infty$ for all regions). Thus, the R_m is an extrapolated measure of the radius of gyration of a hypothetical particle with a *homogeneous* scattering length density and therefore relates to, and is affected by, the overall volume and shape occupied by the combined scattering length density regions. The term α, or the second moment of the internal density fluctuations within the scattering object, is a term that is sensitive to the radial distribution of the scattering length densities of the two regions relative to the object's centre of mass. The β coefficient relates to the displacement between the heterogeneous regions, i.e., the distance between the centre of mass of the particle and the centres of the different of scattering length density regions. Assuming that scattering arising from non-uniform $^1H/^2H$ exchange, incoherent scattering and interparticle interference do not unduly 'warp' the estimation of R_g, then the magnitudes of α and β provide insights into the contrast-weighted disposition between the centres of scattering length density of the individual components within a complex. For example, if $\beta = 0$, the components of a complex share the same centres of scattering length density; a positive α implies the higher scattering density (on average) is located more toward the outside of the particle, while a negative α, places the higher scattering density (on average) more toward the inside of the particle. A zero α implies a homogeneous scattering particle. What the

Stuhrmann relationship elegantly captures is that the R_g is affected by the size and shape of the whole particle as well as its individual components and varies systematically as a function of the contrast depending on the disposition between the heterogeneous regions of scattering length density within a complex. The analysis of the contrast-R_g dependency can be analyzed using the *Rg* module of MULCh [36] (http://smb-research.smb.usyd.edu.au/NCVWeb/).

4 $\Delta\rho = 0$, experimental design and sample preparation.

As mentioned in section 2.4, different classes of biological macromolecules – protein, polynucleotides, lipids, carbohydrates, etc – have different average neutron scattering length densities. Although all atoms have the probability to scatter neutrons, the match points of each macromolecular class are strongly influenced by their average amount of ^{1}H per unit volume, stemming from the very different interactions between neutrons and ^{1}H compared to the other biological isotopes. As a result, $\Delta\rho = 0$ occurs at different percent volume fractions of ^2H_2O in a supporting solvent for each class of biomacromolecules. Two popular experimental approaches are widely used for interrogating macromolecular structures using SANS based on accessing $\Delta\rho = 0$ condition: SANS with contrast matching or SANS with contrast variation. The choice of experiment often boils down to how much sample is available and an understanding what signal is measured at a component match point.

4.1 Contrast matching.

If a macromolecule is covalently bound to, or is in complex with, another molecule that has a different average scattering-length density then the coherent scattering profile obtained at the match point for the first molecule is derived almost-exclusively from the second component. That is, at the match point of macromolecule *x* ($\Delta\rho_x = 0$), it will be possible to obtain shape information from macromolecule *y* and vice versa [57]. This type of experiment is called contrast matching and typically requires measuring the SANS profile of the whole complex (e.g., at 0% v/v ^2H_2O, i.e., in regular light water) and at least one component match point, or preferably two, i.e., where $\Delta\rho_x = 0$ and/or where $\Delta\rho_y = 0$. The % v/v ^2H_2O in the solvent where the expected component and whole-complex match points occur can be calculated *a priori* from the atomic composition of the solvent and the component macromolecules of a complex, for example using the *Contrast* module of MULCh [36]. However, experimentally, when an individual component is selectively matched out at a certain % v/v ^2H_2O in the supporting solvent, the resulting scattering profile – although dominated by the coherent scattering of the bound partner – may still be influenced by additional scattering contributions made by the matched-out component not taken into account by the 'expected' calculations. This includes ^{1}H incoherent scattering and weak scattering arising from ^{1}H/^{2}H solvent-exchange or any other internal particle scattering length density fluctuations. Biological macromolecules are not homogeneous objects and are in dynamic ^{1}H/^{2}H exchange with the solvent and 'unexpected' contributions to the scattering intensities may arise at a component match-point by solvent exchange or internal structural inhomogeneity. These imperfections may become proportionately more influential toward $\Delta\rho = 0$, i.e., where the scattering contribution made to the SANS profile becomes exceedingly weak, but remains nonetheless, and could measurably influence the interpretation of the data.

Differences in scattering between macromolecules of *the same class* may manifest under 'near-but-not-quite $\Delta\rho$=0' conditions, that is, the average scattering length density approximation, which otherwise holds at points away from the match point where the coherent signal is strong, may not hold-up experimentally. For example, proteins may on *average* have a match point in conditions of 43% v/v ^2H_2O, but experimentally and depending on the protein and its exact chemical composition, the match point could be between 40-45% v/v ^2H_2O due to protein-specific internal/localized structure, hydrogen exchange rates/accessibility, etc. If a component is not precisely matched out, then coherent scattering contributions internal to that component and contributions arising from *between* the components of a complex, i.e., the cross-term, could remain. This effect becomes especially acute near a large component match point. As $I(s)$ is proportional to the square of the volume, deviations from the match point of a large component could have quite deleterious effects on the interpretation of a smaller component scattering pattern (eqs.23, 24, and 25). Experimentally, it may be necessary to perform a set of SANS measurements using % v/v ^2H_2O in-and-around the expected match point of a component (e.g., at and +/- 3% v/v ^2H_2O either side of the expected match point) and then apply a careful analysis of $I(0)$, MW and R_g. The advantage of contrast matching is that: 1) it works if the samples are accurately prepared and; 2) consumes less sample compared to a full contrast variation series.

4.2 Contrast variation.

SANS with contrast variation is similar to contrast matching, but typically requires more material and more beam time. For this, contrast variation allows one to extract the scattering functions from individual components of a complex from the set of experimental data, i.e., extract the full scattering curves from the components. A full SANS contrast variation series typically requires preparing both samples and buffers using incremental ratios of ^2H_2O in the solvent. The scattering profiles measured at these contrast points will relate to each other as described in eq.24 under conditions of varying contrast, setting up a set of related expressions. As small-angle scattering is always a combination of sums, the eq.24 relationship derived for each contrast point can be interpolated so that the component scattering functions can be derived from the contrast-set even if the experimental conditions did not exactly satisfy the $\Delta\rho$=0 condition of the individual components, as in, it is not strictly necessary to measure data from the sample at the exact contrast match-points. SANS with contrast variation allows for the deduction of the distance separating the components' centres of contrast (that is reflected in the mass separation between components) as well as the atom pair-distance distribution of the individual components *and* the distance distribution between the individual components of a complex. Therefore, as SANS with contrast variation generates multiple related datasets, the approach effectively 'boosts' the information content of a scattering experiment: The extrapolated component scattering functions and the cross-term yield both the structure and disposition of components within complexes[41] and assemblies [42].

A contrast series can be visualized along the lines using the following example of a 1:1 w/w protein/DNA complex. At 0% v/v ^2H_2O (i.e., pure regular light water) the coherent scattering data will encode shape information about the whole complex, but not the disposition of the individual components within the complex. On increasing the % v/v ^2H_2O in the solvent from 0% v/v up to near the protein match point (near 43 % v/v ^2H_2O in the solvent), the coherent scattering contributions will be increasingly and proportionately dominated by the DNA component until the scattering from the protein component has been almost solvent-matched. Consequently, at the protein match point, the magnitude of the coherent scattering will predominately reflect contributions to the scattering made by the DNA. When increasing the % v/v ^2H_2O in the solvent beyond the protein match point, the coherent scattering signals reduce toward zero until the match point of the *entire* complex is reached. Any registered signal at the entire-complex match point will be very weak and be dominated by incoherent spin-scattering and residual coherent scattering arising from correlated $^1H/^2H$ exchange with the solvent and internal-particle density fluctuations. Increasing the % v/v ^2H_2O beyond the entire-complex match point results in the re-emergence of coherent scattering intensities until at ~60% v/v ^2H_2O, or at the DNA match point, the coherent scattering will be dominated by the protein component alone. Above the DNA match point, DNA scattering contributions begin to add back into the measured profile, reflecting the proportionate scattering contributions from both the protein and DNA components of the complex. Therefore, using SANS with contrast variation, the overall shape of the protein/DNA complex as well as the shape and disposition of its subunits, in principle, can be determined from the coherent scattering data from the whole complex as well as the individual components at their respective % v/v ^2H_2O match points. The *Compost* module of MULCh [36] can be used to solving the set of equations derived from a SANS with contrast variation series (eq.24) and the extraction of the requisite cross-term scattering functions.

4.3 What concentration?

The ability to 'match in' and 'match out' components using SANS influences how samples are prepared [17]. First, there is selecting an appropriate sample concentration. The consequence of contrast variation is that the magnitude of the coherent scattering derived from any one individual component of a complex will be less than the combined scattering intensities derived from both components of the whole complex. In other words, when matching out components, the net SANS intensities will be reduced in relative to the 'start concentration' of the original complex and this will be proportionate to the volume squared of the components. Therefore, it is advisable to begin an experiment at a reasonable sample concentration to ensure that the coherent scattering measured from an individual component is above the background noise derived from the incoherent spin-scattering contributions arising from the solvent. Incoherent scattering contributions can be an issue, especially in the early stages of a contrast variation series (0–50% v/v ^2H_2O) due to the large incoherent-spin scattering length and subsequent cross section of ^1H. But what concentration does one choose? This relates to the volume squared of the whole complex and the volume squared of the individual components that in effect relates to the mass ratios of components. For larger complexes (50 kDa and above e.g., at 5–7 mg.ml^{-1}), incoherent noise from the solvent may not be of concern, but for smaller complexes (50 kDa and below) even at relatively high concentrations (e.g. 6–10 mg ml^{-1}) ^1H incoherent spin-scattering contributions may 'drown out' the coherent scattering. The obvious solution is to simply increase sample concentrations, but then there is a risk of introducing interparticle interference effects. Another option is to increase exposure times (quadrupling collection time should result in a two-fold improvement in counting statistics) but sometimes even this may not be effective, especially if one of the components of a complex is significantly smaller than the other. The solution: separate the match points so that larger components of a complex match in? at high % v/v ^2H_2O to enable data collection from smaller components in a solvent background with low incoherent scattering, i.e., in high % v/v ^2H_2O in the solvent.

4.4 Improving match-point separation using biodeuteration.

Incoherent scattering from the solvent may become increasingly problematic if the mass ratios of the components of a complex become increasingly disproportionate. As the mass ratio tends toward the extreme, not only do the total coherent scattering contributions from a smaller component become proportionately less ($I(0) \propto V^2$), but the separation of the match points between the large component and the entire-complex (where scattering intensities are negligible) become less and less distinguishable. The combined effect complicates the collection of data from the small component. For example, a 75 kDa protein in complex with a 20 kDa piece of DNA will have protein match-point and entire-complex match point that are very close (~42% v/v ^2H_2O and ~45% v/v ^2H_2O). Such a scenario makes it difficult to perform a SANS experiment in terms of accurately making up solvents to the exact % v/v ^2H_2O as well as measuring data above the background noise for the DNA component at the protein match point where incoherent scattering from ^1H at these % v/v ^2H_2O remains significant. The issue of poor match point separation can be solved by altering the ^1H:^2H per unit volume on one of the components, i.e., partially or completely deuterating a component with non-exchangeable ^2H.

The level and extent of ^2H incorporation into a macromolecule can be controlled using biodeuteration, e.g. the expression of a target in *Escherichia coli* B strains in ^2H_2O media[16, 17, 43, 44]. Biodeuteration alters the non-exchangeable ^1H per unit volume on a component and consequently, when a complex is formed between the ^2H-labelled material and an 'all-^1H' binding partner, the contrast against the solvent and between the components of a complex are radically altered. This causes the component and whole complex match points to separate. Match point separation using biodeuteration opens up opportunities to structurally interrogate otherwise difficult macromolecular complexes, such as the analysis of complexes comprised of subunits with disproportionate masses. In the example above, if the protein component of the protein/DNA complex were labelled with ~72% non-exchangeable ^2H, then the match points separate to ~100% (protein), 91% (whole complex) and ~60% (DNA) v/v ^2H_2O, respectively, making it possible to measure the scattering from the protein and the DNA in high % v/v ^2H_2O buffers. Furthermore, and of extreme significance, biodeuteration allows the extraction shape information from ^1H-protein-^2H-protein complexes, that is otherwise almost-impossible to achieve from all-^1H protein systems (Figure 13).

Assessing what average level of non-exchangeable ^2H should be incorporated into a component to obtain decent match point separations within a sample can be predicted *a-priori* using the *Contrast* module of MULCh [36]. The module takes primary protein, RNA or DNA sequences and the atomic composition of the solvent and calculates the match points for a system. It also calculates the contrasts for a hypothetical contrast variation experiment, considering the extent of ^2H-labelling on a component and the percentage of acidic protons likely to be in exchange between a complex and the solvent (usually around 90–95%). These calculations provide an excellent guide for setting up biodeuteration runs (e.g., deciding on what % v/v ^2H_2O to grow cell cultures to obtain the desired level of non-exchangeable ^2H incorporation) and evaluating how SANS signal intensities will change for each component at different contrast values.

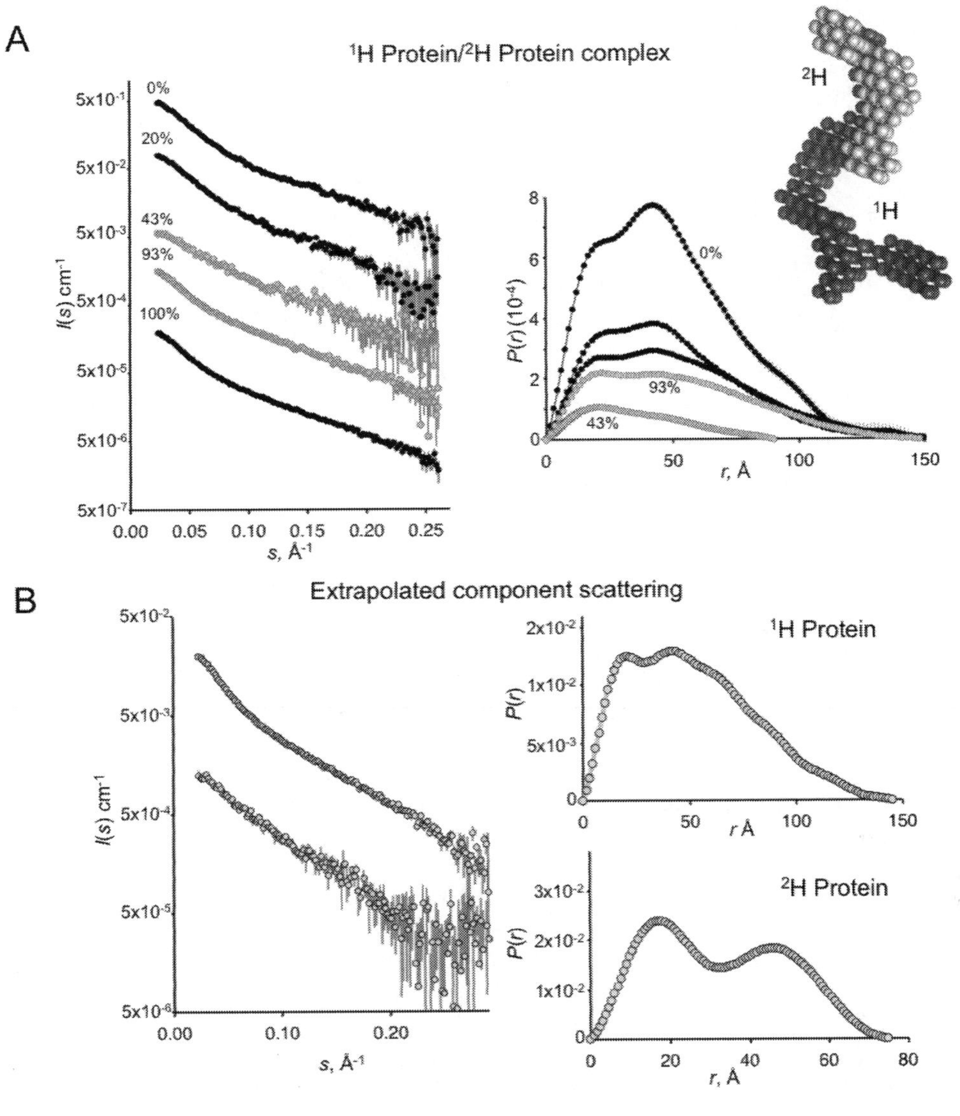

Fig. 13. *A full contrast variation series measured from a ^{1}H and ^{2}H labelled protein-protein complex.* **A.** SANS profiles measured at different %v/v $^{2}H_2O$ in the solvent. The near, but not exact, match point scattering data for the ^{1}H-component (43% v/v $^{2}H_2O$) and the ^{2}H-component (93% v/v $^{2}H_2O$) are coloured grey. The corresponding $p(r)$ profiles obtained each contrast variation point/dataset are shown. **B.** The individual component scattering functions of the ^{1}H- and ^{2}H-protein components extrapolated from the contrast variation data obtained from eq.24 and using the *Compost* module of MULCh, and the corresponding $p(r)$ profiles of each individual protein. The resulting real-space low-resolution representation of the protein-protein complex is shown as a space-filling dummy atom model.

4.5 Is your complex a complex in $^{2}H_2O$?

There are several aspects to consider when preparing samples for a SANS experiment that are provided in detail in [17]. This includes, but is not limited to, assessing sample quality prior to a SANS measurement and the quantities of material required for contrast matching or contrast variation experiments. For example, five or more samples are typically required to span a SANS with contrast variation series that, for a ^{1}H-protein-^{2}H-protein complex, might translate into setting up dialysis against 0%, 20% 43%, 90% 100% v/v ^2H_2O using a total of ~2–2.5 ml of sample in a concentration range somewhere between 5–10 mg.ml^{-1}. The concentration used for the final experiment has to balance between maximizing signal that greatly depends on the V^2 of the components, versus introducing unwanted interparticle interference effects $S(s)$ terms in the scattering at too-high concentration, versus the disassociation constant, Kd, of the components of the complex. What must be reiterated here, is that it is absolutely necessary to evaluate whether samples remain fully associated and soluble in solutions containing high % v/v ^2H_2O. The strength of ^{2}H-hydrogen bonds is different to ^{1}H-hydrogen bonds, as reflected, for example, in the difference between pH and p^{2}H (termed pD, where pD = pH + 0.4). In addition, the solvation layer around a macromolecule has different properties compared to the bulk solvent. The cumulative effects of these differences are that ^2H_2O has the potential to affect the solubility, stability and structural dynamics of macromolecules[45-51] that could result in significant shifts in Kd, resulting in complexes falling apart in ^2H_2O, or, perhaps more commonly, lead to non-specific aggregation. Consequently, it is absolutely required to assess the effects of ^2H_2O on the stability of a sample in order to satisfy conditions where a complex is both stable, soluble and monodisperse as a function ^2H_2O concentration. Indeed, the biggest challenge facing a biomacromolecular SANS experiment, aside from producing sufficient material, is maintaining the integrity of the sample, be it ^{2}H labelled or not, in high % v/v ^2H_2O solutions. As scattering intensity scales to the volume squared, even trace levels of aggregation can ruin SANS experiment, destroying the relationships between the contrast points as described in eq.24 and the subsequent evaluation of accurate structural information.

5 Questions and answers: a student's perspective.

- *What is the first aspect of SANS – be it theory or experiment – that a student of structural biology should be aware of before they begin?*

It is very important to evaluate what pre-existing data is available on your system and to obtain as much information as possible about the properties of the sample, especially in terms of stability, overall size, molecular weight, and the propensity of a sample to aggregate. The aim of the wet lab work is to obtain as pure and monodisperse sample as possible and characterize it with available methods. Beforehand, it is beneficial to perform small angle X-ray scattering (SAXS) experiments as SAXS helps to determine two major factors: i) The optimal sample conditions, e.g., to evaluate the aggregation propensity of the sample and; ii) To estimate the s-range, and in particular the experimental s_{min} required to capture scattering data that spans the longest vector lengths in the sample (as a rule of thumb it is good practice to measure to where $s_{min} = \pi/D_{max}$, or better yet, $s_{min} = 1/D_{max}$). This information is useful to select what sample-detector positions are chosen for the SANS measurements before the experiment takes place. In addition, and what is important for structural biologists, is that SAXS also provides insights into low resolution structure of the sample. For two component complexes, the low-resolution structure obtained from SAXS can be used to complement the SANS data. With the help of neutrons, a more detailed structural picture then emerges as the SANS yields additional information on how the components are oriented inside this low-resolution structure.

The other important consideration is the amount of sample required for SANS. Compared to SAXS (20–50 μl), the quantity of sample required for a full SANS experiment can be a lot more (100s μl to ml) simply because of the size of the sample cells used at a neutron facility (due to the large area of the neutron beam). There are several neutron facilities offering deuteration of biomolecules based on a proposal application system. The use of bio-fermenters results in production of high-density bacterial cell paste with overexpressed ^1H and ^2H-labeled proteins that can really help produce enough material for the experiment compared to lab-based protein over expression. Generally, the least 200–500 μl of a sample at 5–10 mg ml^{-1} are required for a *single* point of SANS contrast series, therefore it might be advantageous to access the resources on offer at biodeuteration facilities.

It is also very important to optimize and characterize samples on a small scale as much as possible before to leaping into a full SANS with contrast variation experiment due to the often heavy sample preparation demands of the experiments. However, one of the advantages of SANS is that (and unlike X-rays) the technique is generally non-destructive which is beneficial even when sample quantity might be limited. In cases where sample amounts did not quite reach planned expectations it is possible, after measuring and checking for radiation activity, to re-use the samples by mixing them together to obtain different contrast points and repeat the procedure with corresponding buffer.

- *What is the most difficult/complicated aspect of SANS for structural biology? How do you overcome these difficulties?*

The most difficult part is optimizing conditions where the samples are, and remain, soluble and stable in high percentage ^2H_2O buffers. The regular buffer composition in ^1H_2O solution might not always be the best for your system when going for 100% v/v ^2H_2O. In some cases, it is necessary to change the base of the buffer, consider using potassium chloride instead/or with sodium chloride, adding deuterated glycerol or a changing to a more stable reducing agent (e.g., tris(2-carboxyethyl)phosphine), or TCEP, instead of dithothreitol, DTT).

Sample mono-dispersity is essential during solution studies and its assessment should be done at different stages of the sample preparation. For this, dynamic light scattering, DLS, is an easy and efficient technique available in most molecular biology laboratories, which allows evaluation of sample quality. There are several things to check:

1) The sample's behavior in high percentage ^2H_2O solutions.
2) Time-induced aggregation, as SANS measurement may requires long time exposures, especially near the component match points.
3) Aggregation formation upon thawing/freezing, which is important to consider when transporting samples to a facility.
4) Sample stability at different temperatures. Depending on the specific beam line, sample holder temperatures below 10°C might not be available.

Of course, and aside from DLS, SAXS may be used to determine the optimal sample conditions and concentrations to assess appearance of interparticle interactions, be they repulsive or attractive (i.e., to assess aggregation).

Furthermore, overexpression of deuterated biomolecules in ^2H minimal media might result in the formation of insoluble inclusion bodies or, in general, lower amounts of soluble protein. It is a good idea to perform a small-scale solubility screen to test different extraction protocols to obtain higher yields of material.

- *Are there aspects of the technique that are easily overlooked/forgotten or misunderstood – theory, instrument, and sample preparation?*

Before setting up a SANS experiment there are several theoretical steps which must be done *in silico*, e.g., using MULCh. It is not only important to obtain the estimate of the components match points, but also to calculate differences in the magnitude of scattering intensities of molecules in different $^{1}H_2O/^{2}H_2O$ ratios so as to help choose what contrast points to measure. The major factor is to decide prior to the experiment on what level of non-exchangeable deuteration is required to obtain the best match point separation of the system components.

Another aspect to consider is the 'wet-lab' perspective on sample preparation and how this may, or may not, be compatible with a SANS instrument/experiment. Typically, the majority of purification steps and sample handling procedures in the wet-lab are performed in the cold room or on ice in order to preserve the sample over the often-extended number of steps and time required to isolate a sample. It is really important to assess the temperature stability of the sample after it has been purified, i.e., can it maintain its integrity for an extended period of time (e.g., up to 2–3 hrs) out of the cold-room? Such information it is important to know. Dilute macromolecule samples are *very*-weakly scattering and although it might be possible to measure SANS at colder-than-room-temperature, colder temperature measurements run the risk of spontaneous water-vapor condensation from the atmosphere on the large sample cells over extended neutron exposure times. Such condensation on the outside of the SANS sample cell can ruin the measurement. So it is important to check what temperature range is available at a beam line and, very importantly, consult with your local beam line facility contact about any difficulties with water condensation at or below 10 °C. There is little point spending time and finding out in the wet-lab that your sample is stable at 4 °C only to discover at the last minute you cannot measure it at 4 °C.

When travelling to a facility it is a good idea to devise an experimental plan and to take some of your own consumables, for example, pack your own pipettes and tips. It can take a while to familiarize yourself with a facility user lab and to figure out where everything is. Having some of your own consumables at hand simply relieves the stress of running around looking for particular items. In regards to the experimental plan, have a well-prepared/thought-out and systematic dialysis protocol, that limits the number of steps required to complete the dialysis and, hopefully, the stress involved with the sample handling. Write the protocol down, having already pre-established the quantity of buffer components required, sample concentrations, etc, prior to setting up the dialysis at the facility. A written protocol helps set a routine and clarifies what to do at 4 am in the morning after having worked for 15 hrs straight setting up the samples. It is also easier to travel with team members, where members have been allocated into 'instrument responsibles' – who learn how to load and drive the instrument – and others as 'sample responsibles' – who deliver the samples to the instrument.

Considerations for the dialysis of samples at a facility include: i) Take your own dialysis equipment; ii) When dialyzing against different $^{2}H_2O$ solutions it is crucial to avoid formation of air pockets in the sample to ensure proper proton-deuterium exchange and inadvertent changes in sample concentration; iii) Even when invisible to a naked eye, $^{2}H_2O$ buffers, especially cold buffers, tend to release bubbles over time (especially when warming up) due to dissolved gasses. These microbubbles scatter like crazy. It may take a few hours to fully remove them by sonication or degassing equipment. Dialyzing at room temperature, or warming up the samples to room temperature in advance of the SANS beam time, limits the dissolved-gas problem; iv) Always centrifuge the post dialysis samples and buffers at full speed in a bench-top centrifuge (e.g., 30 000 x g) to help get rid of bubbles and potential small insoluble or large soluble aggregates. Do this immediately prior to loading into the SANS sample cuvette; v) It is good practice to measure accurate concentrations of both the pre- and post-dialysis samples and the post-measured samples (after passing radiation safety). These concentration measurements may indicate if something unwanted happened during dialysis i.e. aggregates formation or precipitation, or during the measurement; vi) If possible, apply for simultaneous access to synchrotron SAXS beam line or bench-top SAXS instrument during your SANS experiment to get additional data from the post-SANS sample.

- *In what situations do you think SANS is useful for structural biology investigations, that is, what types of questions do you ask that SANS can answer that other techniques cannot?*

High-resolution structure determination techniques are the key to revel the atomic details of biomolecular interactions. However, the beauty of implementing SANS is that it can be used to analyze almost-any biological system in solution covering wide molecular-weight range across almost-any sample environment. With new advances in production of deuterium labelled biological components, SANS can be used to measure very complex systems inaccessible to many standard structural biology techniques, including membrane proteins, in more biologically relevant environment(s). Additionally, SANS is an exciting tool to probe conformational changes caused by interactions, e.g., upon binding of intrinsically disordered proteins. Structural flexibility or intrinsic disorder of macromolecules is not a limitation. The other advantage of SANS is to push structural biology investigations beyond the more traditional 'structures of monodispersed systems' type of experiment, e.g., to ask questions relating to fibrillation, aggregation, gelation, the formation of ordered assemblies, etc. Finally, and just point out, neutrons are relatively non-destructive compared to X-rays so are great for studying samples that are highly susceptible to X-ray radiation damage.

6 Summary.

Small angle neutron scattering affords a unique method to interrogate the size, structure and spatial disposition of macromolecules, complexes, assemblies, nanoconjugates, etc, due to the unique way neutrons scatter from ^{1}H compared to other biological isotopes. The ability to control the coherent neutron scattering contrast in such systems via the isotopic substitution of ^{1}H with ^{2}H affords a way of assessing the internal structure and arrangement of individual component in higher-order systems. SANS is not confined to a specific sample environment, allowing one to access diverse experimental samples under diverse experimental conditions including particles in near-native solutions, flexible macromolecules, particles under shear, pressure or extremes in temperature and pH, phase-changes, etc. This chapter outlines the most basic explanation of SANS that will hopefully be of use for structural biology students and, therefore, helping structural biology going forward.

Acknowledgements

The authors wish to acknowledge the European Union's Horizon 2020 research and innovation programme 'iNEXT Discovery' grant agreement No 871037. ZP is funded by the the Swedish Foundation for Strategic Research (SSF) within the Swedish national graduate school in neutron scattering (SwedNess). We would also like to thank Patrice Vachette of the Institut de Biologie Intégrative de la Cellule, Université Paris-Sud (for general inspiration) as well as the Société Française de la Neutronique, including the travel support/resources made available during the 2018 'Neutrons and Biology' School, Carqueiranne, France.

References

1. Z.H. Zhou, Adv. Protein Chem. Struct. Biol., **82**, 1 (2011)
2. M.A. Herzik, Jr., M. Wu, G.C. Lander, Nat. Commun., **10**, 1032 (2019)
3. M.G. Campbell, et al., Structure, **20**, 1823 (2012)
4. A.F. Brilot, et al., J. Struct. Biol., **177**, 630 (2012)

5. J.R. Helliwell, Acta. Crystallogr. D Struct. Biol., **76**, 87 (2020)
6. T. Sugiki, N. Kobayashi, T. Fujiwara, Comput. Struct. Biotechnol. J., **15**, 328 (2017)
7. J.K. Krueger and G.D. Wignall, *Small-Angle Neutron Scattering from Biological Molecules*, in *Neutron Scattering in Biology*. 2006, Springer Berlin Heidelberg. p. 127.
8. J.H. Lakey, Journal of the Royal Society, Interface / the Royal Society, **6 Suppl 5**, S567 (2009)
9. W.T. Heller and K.C. Littrell, Methods Mol. Biol., **544**, 293 (2009)
10. S. Krueger, Physica B: Condensed Matter, **241–243**, 1131 (1997)
11. E. Mahieu and F. Gabel, Acta Crystallogr D Struct Biol, **74**, 715 (2018)
12. C. Breyton, et al., Eur. Phys. J. E Soft Matter, **36**, 71 (2013)
13. D.I. Svergun, et al., *Small Angle X-Ray and Neutron Scattering from Solutions of Biological Macromolecules*. 1 ed. 2013: Oxford University Press.
14. *Neutron scattering in biology techniques and applications*, ed. J. Fitter, T. Gutberlet, and J. Katsaras. 2006, Berlin Heidelberg: Springer-Verlag.
15. G. Zaccai, Eur. Biophys. J., **41**, 781 (2012)
16. N.R. Zaccai, et al., Methods Enzymol., **566**, 159 (2016)
17. C.M. Jeffries, et al., Nat. Protoc., **11**, 2122 (2016)
18. A.D. Martin, et al., Sci Rep, **7**, 43947 (2017)
19. M. Martínez-Sanz, M.J. Gidley, E.P. Gilbert, Soft Matter, **12**, 1534 (2016)
20. P. J.K. and M.J.A. Hore, Curr. Opin. Colloid Interface Sci., **42**, 157 (2019)
21. I. Yadav, et al., Langmuir, **33**, 1227 (2017)
22. R.F. Tabor, J. Eastoe, I. Grillo, Soft Matter, **5**, 2125 (2009)
23. J. Siefker, et al., J. Am. Chem. Soc., **140**, 12720 (2018)
24. B. E., et al., J. App. Cryst, **42**, 461 (2009)
25. A. Guinier, Ann. Phys. (Paris), **12**, 161 (1939)
26. A. Guinier and G. Fournet, *Small Angle Scattering of X-Rays*. 1955, New York: Wiley.
27. D.I. Svergun, J. Appl. Crystallogr., **25**, 495 (1992)
28. O. Glatter, J. Appl. Cryst., **10**, 415 (1977)
29. L.A. Feigin and D.I. Svergun, *Structure analysis by small-angle x-ray and neutron scattering*. 1987, New York: Plenum Press. xiii.
30. D.I. Svergun and K.H. Nierhaus, J. Biol. Chem., **275**, 14432 (2000)
31. G.L. Squires, *Introdution to the theory of thermal neutron scatternig*. 1996, Mineola, New York: Cambridge University Press.
32. J. Chadwick, Proceedings of the Royal Society A, **136**, 692 (1932)
33. L. Meitner and O.R. Frisch, Nature, **143** 239 (1939)
34. O. Hahn and F. Strassmann, Naturwissenschaften, **27**, 11 (1939)
35. V.F. Sears, Neutron News, **3**, 26 (1992)
36. A.E. Whitten, S. Cai, J. Trewhella, J. Appl. Crystallogr., **41**, 222 (2008)
37. S. Maric, et al., Acta. Crystallogr. D Struct. Biol., **70**, 317 (2014)
38. T.E.t. Cleveland, et al., Protein science, **27**, 780 (2018)
39. Z. Luo, et al., Nat. Commun., **9**, 1343 (2018)
40. M. Moglianetti, et al., Chem. Sci., **5**, 1232 (2014)
41. A.E. Whitten, et al., J. Mol. Biol, **368**, 407 (2007)
42. A.E. Whitten, et al., Proc. Natl. Acad. Sci. USA, **105**, 18360 (2008)
43. B. Leiting, F. Marsilio, J.F. O'Connell, Anal. Biochem, **265**, 351 (1998)
44. X. Chen, et al., Food and Bioproducts Processing, **90**, 563 (2012)
45. P.A. Baghurst, L.W. Nichol, W.H. Sawyer, J. Biol. Chem., **247**, 3199 (1972)
46. L. Dougan, et al., Chemphyschem, **9**, 2836 (2008)
47. M. Tehei, et al., Proc. Natl. Acad. Sci. USA, **98**, 14356 (2001)

48. M. Jasnin, et al., Eur. Biophys. J., **37**, 613 (2008)
49. D.I. Svergun, et al., Proc. Natl. Acad. Sci. USA., **95**, 2267 (1998)
50. A.S. Goryunov, Gen. Physiol. Biophys., **25**, 303 (2006)
51. S.Y. Sheu, et al., J. Phys. Chem. A, **112**, 797 (2008)
52. H.B. Stuhrmann, J. App. Cryst., **7**,173 (1974)
53. H.B. Stuhrmann, A. Miller, J. App. Cryst., **11**, 325 (1978)
54. J. Trewhella et al., Acta Crystallogr D, **73**, 710 (2017)
55. R. Dos Santos Morais, et al., Langmuir, **33**, 6572 (2017)
56. R. Dos Santos Morais et al., Biophys. J., **115**, 1231 (2018)

The power of SANS, combined with deuteration and contrast variation, for structural studies of functional and dynamic biomacromolecular systems in solution

Emilie Mahieu[1], *Ziad* Ibrahim[1,2,#], *Martine* Moulin[2], *Michael* Härtlein[2], *Bruno* Franzetti[1], *Anne* Martel[2], and *Frank* Gabel[1,*]

[1]Univ. Grenoble Alpes, CEA, CNRS, IBS, 38044 Grenoble, France

[2]Institut Laue-Langevin, 38044 Grenoble, France

[#]*Present address*: Leicester Institute of Structural and Chemical Biology, University of Leicester, United Kingdom

Abstract. Small-angle neutron scattering (SANS), combined with macromolecular deuteration and solvent contrast variation (H_2O/D_2O exchange) allows focussing selectively on the signal of specific proteins in multi-protein complexes or mixtures of isolated proteins. We illustrate this unique capacity by the example of a functional protein-degradation system in solution, the PAN-20S proteasome complex in the presence of a protein substrate, ssrA-tagged GFP. By comparing experimental SANS data with synthetic SAXS (small-angle X-ray scattering) data, predicted for the same system under identical conditions, we show that SANS, when combined with macromolecular deuteration and solvent contrast variation, can specifically focus on the conformation of the PAN unfoldase, even in the presence of very large GFP aggregates. Likewise, structural information of native GFP states can be visualized in detail, even in the presence of the much larger PAN-20S unfoldase-protease oligomers, which would dominate the overall scattering signal when using X-rays instead of neutrons.

1 Introduction

Small-angle X-ray (SAXS) and neutron (SANS) scattering provide structural information on biomacromolecules in solution on the nanometre length scale (*1*). While X-rays are scattered by the electrons in the macromolecules and the buffer, neutrons are scattered by the respective atomic nuclei. The SAS (small-angle scattering) signal of a solubilized biomacromolecule is a Fourier transform, averaged over all possible orientations, of its X-ray (neutron) scattering length density (SLD) difference with respect to the solvent. The signal interpreted is usually an azimuthally averaged 1D curve (intensity vs. scattering angle or q-value), that informs about the molecular weight of a solubilized particle (via the intensity

* Corresponding author: frank.gabel@ibs.fr

scattered at zero angle, I(0)) and its linear dimensions (via the radius of gyration, R_G). Low-resolution shapes can be generated from the 1D scattering curves or, if atomic-resolution models are available, their back-calculated SAXS/SANS curve be compared with the experimental data (*2*).

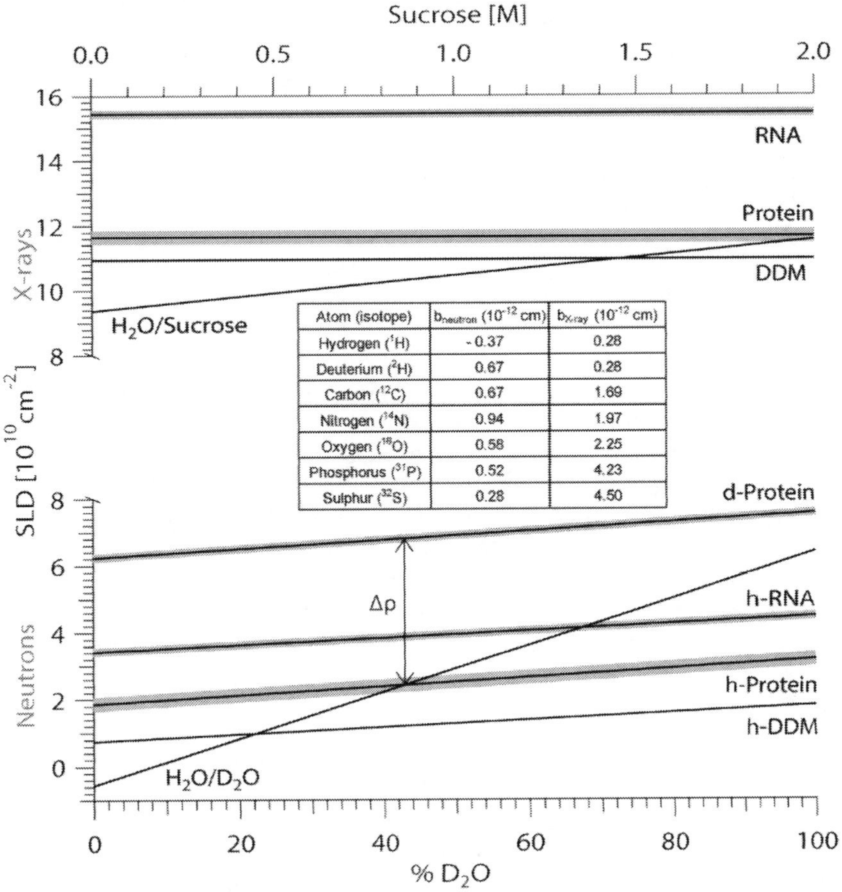

Atom (isotope)	$b_{neutron}$ (10^{-12} cm)	$b_{X\text{-}ray}$ (10^{-12} cm)
Hydrogen (^{1}H)	- 0.37	0.28
Deuterium (^{2}H)	0.67	0.28
Carbon (^{12}C)	0.67	1.69
Nitrogen (^{14}N)	0.94	1.97
Oxygen (^{18}O)	0.58	2.25
Phosphorus (^{31}P)	0.52	4.23
Sulphur (^{32}S)	0.28	4.50

Figure 1. Scattering length densities (SLD) of neutrons and X-rays for several biomolecules. (adopted after (*3*)). Δρ indicates the contrast of a d-protein at 42% D_2O where the h-protein is matched. *Inset*: Neutron and X-ray scattering lengths for major light atoms found in biomolecules. DDM (n-dodecyl-β-D-maltopyranoside) is shown as an example of detergent used to solubilize membrane proteins. SLDs of DNA are not shown explicitly but are close to those of RNA. The deuteration state of the molecules is not indicated for the SAXS SLDs since X-rays do not distinguish hydrogenated and deuterated molecules.

Since the X-ray scattering length of an atom depends linearly on the number of electrons, X-rays are particularly sensitive to heavier atoms and DNA/RNA molecules scatter, per unit volume, stronger than a protein in an aqueous buffer, due to their sugar-phosphate backbones. It is therefore in principle possible for SAXS to distinguish proteins from RNA/DNA (and lipids/detergents in the case of solubilized membrane proteins) by modifying the solvent electron density by the addition of small, electron-rich molecules such as sugar (Fig. 1). Even though there have been some recent advances using medical contrast media (*4*), classical contrast agents such as sugar reach, in general, hardly the electron density of proteins (*5*). They can therefore not match the electron density of RNA/DNA in order to mask their scattering contribution and to focus on the signal of proteins in a protein-RNA/DNA complex. Moreover, the high (molar) concentrations of most contrast agents pose problems of solubility and integrity of many biomacromolecular systems and practical issues due to high viscosities (*4*). Finally, SAXS cannot distinguish between different proteins in solution since they have, in general, a comparable average electron density and contrast with the solvent, i.e. their intrinsic electron density cannot be chemically modified in a global and homogeneous way. SAXS data on multi-protein complexes therefore represents an overall envelope (or shape) of the complex but does not, in general, reveal the internal arrangement and conformations of the protein partners.

In contrast to X-rays, neutron scattering lengths depend in an irregular way on the atomic number, but also on the isotope state of a given atom ((*1*), inset Fig. 1). The most important difference between neutron scattering lengths in biological samples is the one between protons (^1H) and its isotope deuterium (^2H or D). Indeed, a continuous change from a 100% H_2O buffer solution to a 100% D_2O (heavy water) solution allows to cover the entire range of natural (i.e. hydrogenated) SLDs of all relevant biomolecules (including proteins, detergent/lipids, and RNA/DNA) (Fig. 1). Even though the exchange of hydrogen against deuterium in the solvent increases the tendency of certain macromolecules to aggregate, it is in general more "gentle" than the one with small electron-rich molecules in SAXS.

Moreover, selective deuteration of bio(macro)molecules (*6*) allows to increase the natural contrast between chemically different molecules, or create an artificial contrast between chemically similar molecules. **In this way, it is possible to separate the signal from protonated/hydrogenated (h) and deuterated (d) proteins in a complex, and focus on the structural signal from either of them separately. E.g., at about 42% D_2O, the buffer SLD is identical to that of hydrogenated protein, while a perdeuterated (i.e. fully deuterated) protein has a strong contrast** (Fig. 1). Finally, neutrons do not induce radiation damage (as do X-rays) and biological samples can be exposed for several hours and remain biochemically active (*7*).

Here, we compare experimental SANS data with synthetic SAXS data, calculated for the same experimental conditions, to illustrate the unique capacity of SANS to focus on the structural information of specific molecules in a mixture of several protein species of extremely variable size, shape, and oligomeric state.

As a concrete example, we use SANS data from the PAN-20S proteasome system (Fig. 6), in the presence of an ssrA-tagged GFP which is recognized by PAN-20 as a substrate (*8*). The proteasome is a large, hetero-oligomeric macromolecular complex, responsible for specific, ATP-driven degradation of dysfunctional, misfolded, or superfluous proteins in the cell (*9*). In the case of the hyperthermophilic archeon *Methanocaldococcus jannaschii*, it consists of a hexameric, 298 kDa unfoldase PAN, which recognizes and unfolds substrates, and transfers them into the protease 20S particle (751 kDa) where substrates are proteolyzed into small oligopeptide products. We used this robust and thermostable archaeal system to degrade a Green Fluorescent Protein (GFP) variant (28 kDa) containing an ssrA-tag, which allows PAN to recognize it as a substrate (*10, 11*). By using selective per-deuteration (i.e. full deuteration) of either PAN or GFP in a 42% D_2O buffer, we were able to focus selectively on either of both proteins during the active degradation process: under these labelling and solvent contrast conditions, the signal of the hydrogenated protein species in a reaction mixture are completely masked and the perdeuterated one has a strong contrast (Fig. 1). We demonstrate that this is even true if the hydrogenated species or oligomerization states are several orders of magnitude bigger than the deuterated ones. By thermo-activation at 55 °C, the protein degradation reactions were followed at a time-resolution of 30 seconds during 45 minutes, coupled with online fluorescence spectroscopy (*8*).

2 Material and methods

Synthetic SAXS data were calculated for native GFP (*8*), PAN (PDB ID 6HEA), 20S (PDB IB 6HEA), and a GFP aggregate by using the program *IMSIM* from the ATSAS 3.0.0 suite (*12*). Theoretical intensities, including errors, were theoretically calculated for 1 second exposure times on a Pilatus 6M detector at the P12 SAXS beamline at PETRAIII for protein concentrations as indicated in the legends of Figures 2-7, corresponding to the ones in the respective SANS experiments.

Experimental SANS data were recorded on D22 at the European Neutron Source Institut Laue-Langevin (Grenoble, France) on selectively perdeuterated (d) GFP or PAN, following sample preparation and deuteration protocols described elsewhere (*8*).

3 Results

3.1 The PAN-GFP system in the absence of 20S

Isolated PAN unfolds ssrA-tagged GFP when ATP is supplied (*13*). In the absence of the proteolytic 20S partner, the released unfolded products instantly form large-scale aggregates in solution (*8*). Is it possible to extract structural information on PAN and follow potential conformational changes during ATP-hydrolysis in the presence of large-scale GFP aggregates? Figure 2 shows the expected SAXS curves of the individual partners as well as the overall signal, simulated for a 1 second exposure time on a high-performance SAXS synchrotron beamline. The overall signal (black) is completely dominated by the GFP aggregates and cannot be interpreted in terms of PAN structures or conformational changes. Indeed, the I(0) intensities of the aggregates are more than 30 times stronger (please note the logarithmic scale) than those of PAN for the given concentrations.

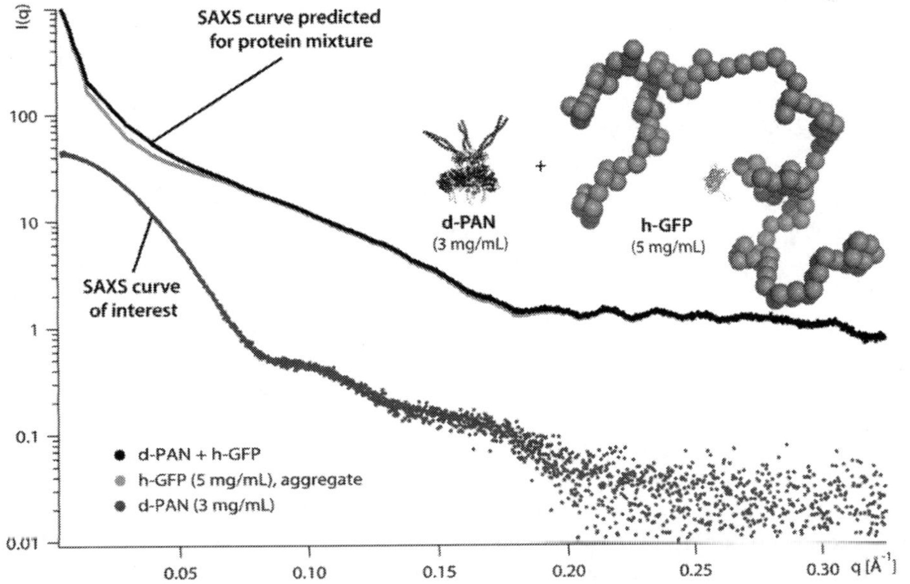

Figure 2. SAXS curve predicted for the d-PAN + h-GFP reaction mixture. The SAXS curve of the reaction mixture (black) is the sum of both species, and is strongly dominated by the h-GFP aggregates (light grey). The structural information on d-PAN (dark grey) is not accessible.

Experimental SANS data (in a 42% D_2O buffer) from perdeuterated (d) PAN and protonated (h) GFP, recorded at the same concentrations as those imposed for the generation of the synthetic SAXS data (Fig. 2), are shown in Figure 3. Impressively, the signal from the very large GFP aggregates is almost completely masked and the overall signal from the mixture can be attributed almost exclusively to PAN. Indeed, in the previous original study, the change of this signal, over an experimental time of 45 minutes, was interpreted in terms of conformational changes of PAN due to the hydrolysis of ATP and the unfolding of GFP (*8*).

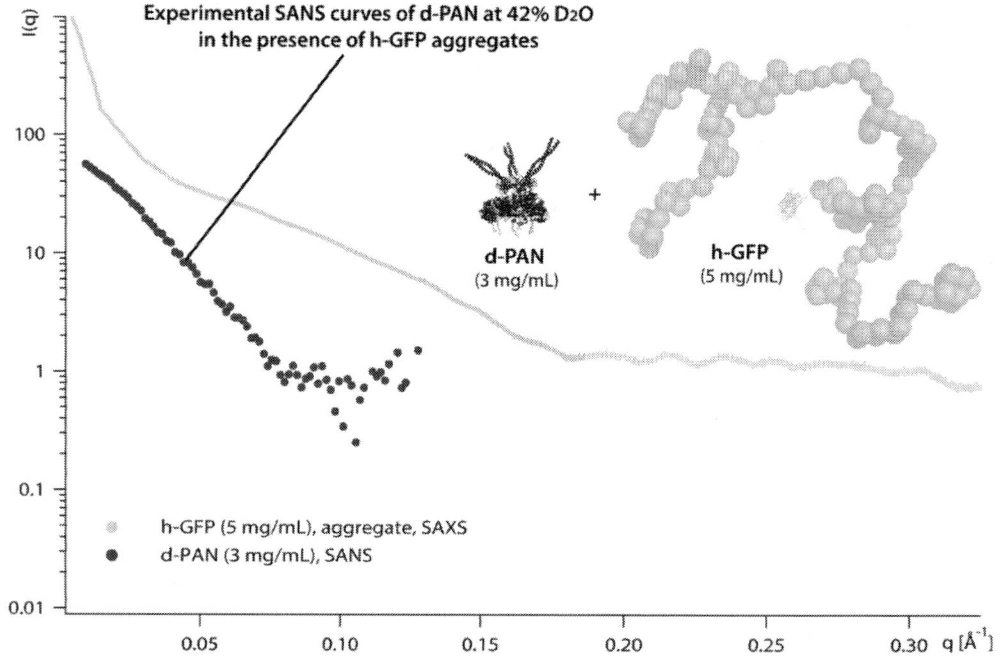

Figure 3. Experimental SANS curve of the d-PAN + h-GFP reaction mixture. In 42% D_2O, the signal of the h-GFP aggregates (transparent grey) is matched and the final SANS curve (dark grey) can be attributed almost exclusively to d-PAN.

Is it possible, on the other hand, to interpret the signal of a mixture of GFP and PAN only in terms of GFP aggregates by neglecting the PAN scattering intensity? Figure 4 shows the expected SAXS curves of such a scenario. While GFP aggregates dominate the overall SAXS signal predicted for the mixture (black data points), there is a non-negligible contribution from PAN, especially at low angles ($q < 0.07$ Å^{-1}) which, if neglected, would lead to an underestimation of the size and extension of the GFP aggregates and SAXS, again, does not yield the true structural values of the aggregates.

If the same reaction mixture and concentrations are measured by SANS in 42% D_2O on h-PAN and d-GFP on the other hand, the signal from PAN (light grey) is matched out and the remaining signal can be attributed exclusively to the GFP states during the reaction (Fig. 5). It is therefore possible to follow the continuous evolution from native GFP at the beginning of the reaction (grey) until the build-up of the final, large aggregates (dark grey) at the end of neutron exposure after 45 min at 55 °C. Both the small native GFP molecules at the beginning of the reaction as well as the large-scale aggregates towards the end of the reaction are faithfully represented by the overall SANS curves and it is possible to analyse and interpret their evolution as a function of time (*8*).

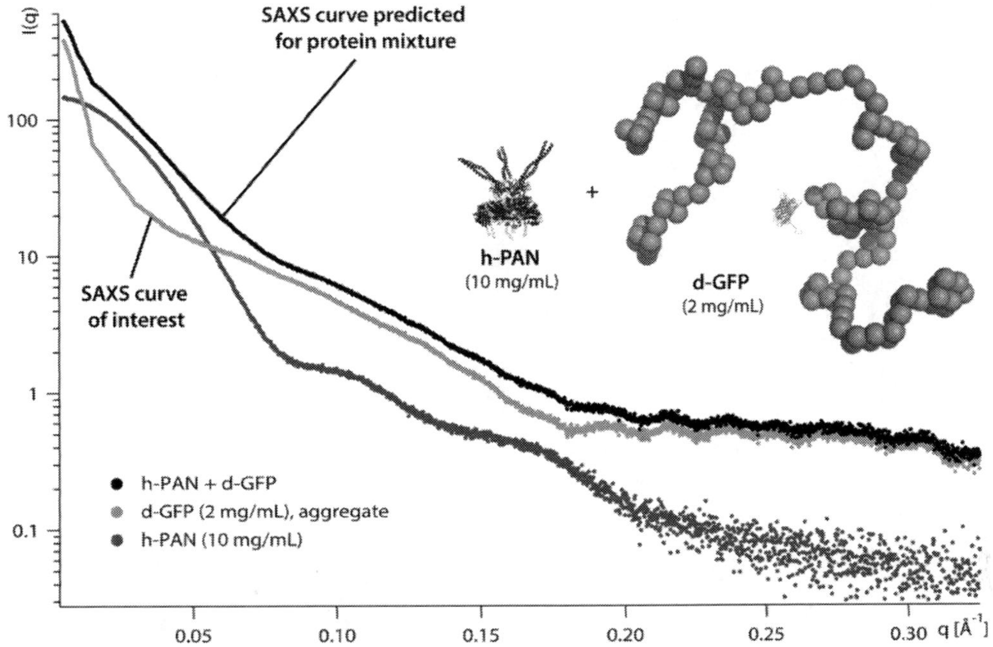

Figure 4. Predicted SAXS curve of the h-PAN + d-GFP reaction mixture. The SAXS curve of the reaction mixture (black) is the sum of the curves of both proteins. While it is dominated by the d-GFP aggregates (light grey), it is contaminated by the signal of d-PAN (dark grey) and can therefore not be interpreted exclusively in terms of the d-GFP aggregates.

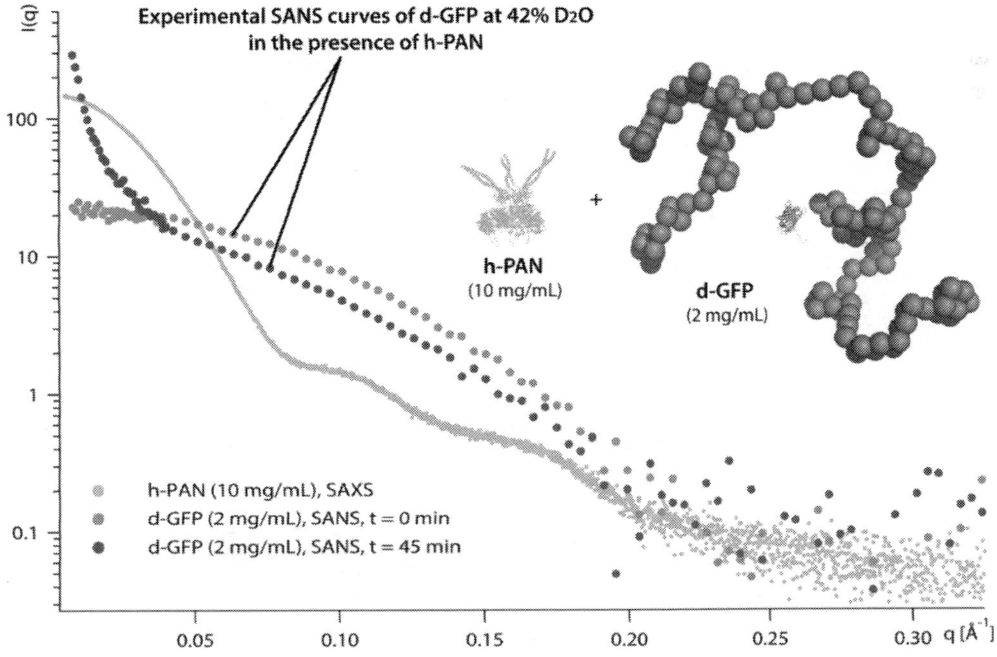

Figure 5. Experimental SANS curve of the h-PAN + d-GFP reaction mixture. In 42% D_2O, the signal of h-PAN (transparent grey) is matched and the SANS curves (grey and dark grey) can be attributed exclusively to native d-GFP (t = 0 min) and d-GFP aggregates formed (t = 45 min).

3.2 The fully functional PAN-20S-GFP system

When a PAN solution is prepared in the presence of 20S and supplied with ATP, GFP molecules are efficiently proteolyzed into small oligopeptide products [unpublished data]. In this more complex, ternary system, how are the chances to observe the relatively small (28 kDa) GFP substrate in the presence of the very large oligomeric PAN and 20S particles (298 and 751 kDa, respectively)? Figure 6 shows the expected overall SAXS signal (black) from a mixture of all three partners, as well as the individual SAXS curves simulated for all molecules at the respective experimental SANS concentrations. As can be seen, the overall SAXS signal is largely dominated by, and representative of, 20S. PAN and GFP contributions are negligible in the global SAXS signal. Therefore, it would be very difficult to interpret the measured 1D SAXS curve of the mixture in terms of native GFP structural states or states of similar size.

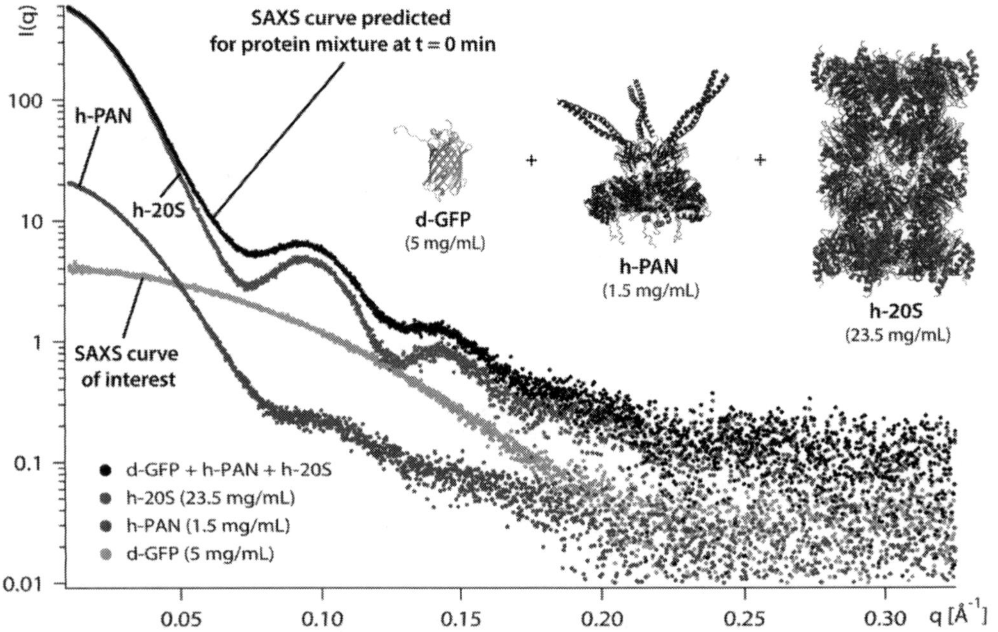

Figure 6. Predicted SAXS curve of the d-GFP + h-PAN + h-20S reaction mixture. The SAXS curve of the reaction mixture (black) is the sum of the curves of all three proteins. It is strongly dominated by the signal of h-20S and it is not possible to extract structural information on d-GFP (light grey).

What would SANS data, measured on the same d-GFP/h-PAN/h-20S system with identical individual concentrations, look like when the hydrogenated, large partners PAN and 20S are masked in 42% D_2O? Figure 7 shows such experimental SANS data in 42% D_2O, measured at concentrations equivalent to those of the synthetic SAXS data in Figure 6. Very impressively, the overall experimental SANS curve represents the structure of native GFP molecules, and the signals of the much larger PAN and 20S particles are very efficiently masked. It is therefore possible, by SANS, in combination with selective deuteration and solvent contrast variation, to follow the state of the small GFP protein substrate during its degradation process by PAN and 20S, an achievement which would be completely impossible to accomplish by SAXS.

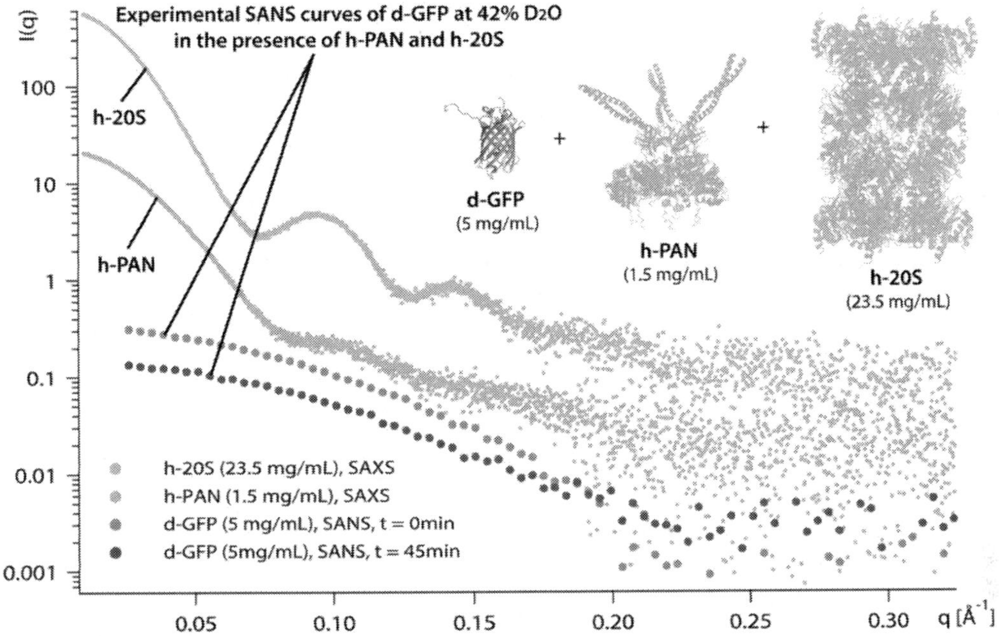

Figure 7. Experimental SANS curve of the d-GFP + h-PAN + h-20S reaction mixture. In 42% D_2O, the strong signals of h-PAN and h-20S (both transparent grey) are matched out and the SANS curves (grey and dark grey) can be attributed exclusively to d-GFP derivatives (native proteins and oligomeric products).

4 General comparison of SAXS and SANS

Both SAXS and SANS allow to obtain structural information (at about the nanometer scale) on solubilized biomacromolecules and the complexes they form. Either technique has its intrinsic strengths and weaknesses:

1) The SAXS technique is more readily available and accessible through a broad geographical distribution of SAXS beamlines at synchrotrons worldwide but also due to widely spread benchtop SAXS instruments, fabricated by a number of companies, and available at various research facilities. Sample amounts required at synchrotron beamlines are modest (~5-50 µL at ~mg/ml concentrations) and exposure times are short (~ms to ~s). No labelling (e.g. deuteration) is required for biomolecules, but samples must be monodisperse if a sophisticated structural interpretation in terms of single species is the aim of a study (*14*). Many biomacromolecules are sensitive to radiation damage by X-rays but protocols have been developed to quantify and deal with this effect (*15*). Generally, SAXS is a quick, and even high-throughput, method to obtain model-free structural parameters (molecular mass, radius of gyration), but also low-resolution envelopes of isolated biomacromolecules, complexes or mixtures of several species (*1*). SAXS is in general not very efficient in separating the structural information from individual partners in an assembled complex (protein-protein, protein-RNA/DNA, solubilized membrane protein). However, atomic models can be scored against SAXS data up to medium/high q-values (~0.5 $Å^{-1}$) with good efficiency and minor conformational changes can be detected.
2) SANS is a "low-throughput" method requiring larger sample amounts (~200 µL at ~mg/mL concentrations), H_2O-D_2O buffer exchange, and, often, the specific deuteration of bio(macro)molecules (*3*). Exposure times are of the order of a minute

up to hours, as a function of contrast, incoherent buffer background, sample concentration and molecular weight. However, neutrons do not induce radiation damage and samples can be modified (buffer exchange, titration of ligands/partners etc) for multiple use in an experimental series. Most importantly, as demonstrated by the data presented here, SANS allows to specifically extract structural information from selected macromolecules in a mixture, even if they are chemically similar (i.e. different proteins). SANS is therefore a technique that requires more investment in sample preparation and yields lower throughput. However, it allows to access and separate structural information of specific isolated macromolecules in a mixture or of partners within an assembled complex, even in the presence of much larger species that would completely dominate the signal if the sample was measured by SAXS.

Acknowledgements

We acknowledge financial support from the French Agence Nationale de la Recherche (grant "PROTStretch" ANR-15-CE11-0026-01). This work used the platforms of the Grenoble Instruct-ERIC Centre (ISBG: UMS 3518 CNRS-CEA-UGA-EMBL) with support from FRISBI (ANR-10-INBS-05-02) and GRAL (ANR-10-LABX-49-01) within the Grenoble Partnership for Structural Biology (PSB). The authors thank Daniel Franke (EMBL Hamburg) for providing a script for the calculation of synthetic SAXS curves with *IMSIM*. The SANS raw data are available under the following DOIs: https://doi.ill.fr/10.5291/ILL-DATA.8-03-935 and https://doi.ill.fr/10.5291/ILL-DATA.8-03-836.

References

1. D. I. Svergun, M. H. J. Koch, P. A. Timmins, R. P. May, *Small Angle X-ray and Neutron Scattering from Solutions of Biological Macromolecules*. I. U. o. Crystallography, Ed., IUCr Monographs on Crystallography (Oxford University Press, Oxford, 2013).
2. C. D. Putnam, M. Hammel, G. L. Hura, J. A. Tainer, X-ray solution scattering (SAXS) combined with crystallography and computation: defining accurate macromolecular structures, conformations and assemblies in solution. *Quarterly Reviews of Biophysics* **40**, 191-285 (2007).
3. E. Mahieu, F. Gabel, Biological small-angle neutron scattering: recent results and development. *Acta Crystallographica Section D* **74**, 715-726 (2018).
4. F. Gabel, S. Engilberge, J. Pérez, E. Girard, Medical contrast media as possible tools for SAXS contrast variation. *IUCrJ* **6**, 521-525 (2019).
5. C. M. Jeffries *et al.*, Preparing monodisperse macromolecular samples for successful biological small-angle X-ray and neutron-scattering experiments. *Nat Protoc* **11**, 2122-2153 (2016).
6. M. Haertlein *et al.*, in *Methods in Enzymology,* Z. Kelman, Ed. (Academic Press, 2016), vol. 566, pp. 113-157.
7. M. Jasnin, M. Moulin, M. Haertlein, G. Zaccai, M. Tehei, Down to atomic-scale intracellular water dynamics. *EMBO Rep* **9**, 543-547 (2008).
8. Z. Ibrahim *et al.*, Time-resolved neutron scattering provides new insight into protein substrate processing by a AAA+ unfoldase. *Scientific Reports* **7**, 40948 (2017).
9. A. L. Goldberg, Protein degradation and protection against misfolded or damaged proteins. *Nature* **426**, 895-899 (2003).
10. E. U. Weber-Ban, B. G. Reid, A. D. Miranker, A. L. Horwich, Global unfolding of a substrate protein by the Hsp100 chaperone ClpA. *Nature* **401**, 90-93 (1999).

11. S. Gottesman, E. Roche, Y. Zhou, R. T. Sauer, The ClpXP and ClpAP proteases degrade proteins with carboxy-terminal peptide tails added by the SsrA-tagging system. *Genes & development* **12**, 1338-1347 (1998).
12. D. Franke *et al.*, ATSAS 2.8: a comprehensive data analysis suite for small-angle scattering from macromolecular solutions. *Journal of Applied Crystallography* **50**, 1212-1225 (2017).
13. N. Benaroudj, P. Zwickl, E. Seemüller, W. Baumeister, A. L. Goldberg, ATP hydrolysis by the proteasome regulatory complex PAN serves multiple functions in protein degradation. *Molecular Cell* **11**, 69-78 (2003).
14. J. Trewhella *et al.*, 2017 publication guidelines for structural modelling of small-angle scattering data from biomolecules in solution: an update. *Acta Crystallographica Section D* **73**, 710-728 (2017).
15. J. C. Brooks-Bartlett *et al.*, Development of tools to automate quantitative analysis of radiation damage in SAXS experiments. *Journal of Synchrotron Radiation* **24**, 63-72 (2017).

Solution structure of macromolecules using small angle neutron scattering and molecular simulations

Jayesh S. Bhatt[1,*]

[1]University College London, Gower Street, London WC1E 6BT, UK

Abstract. An introductory account of using molecular simulations to deduce solution structure of macromolecules using small angle neutron scattering data is presented for biologists. The presence of a liquid solution provides mobility to the molecules, making it difficult to pin down their structure. Here a simple introduction to molecular dynamics and Monte Carlo techniques is followed by a recipe to use the output of the simulations along with the scattering data in order to infer the structure of macromolecules when they are placed in a liquid solution. Some practical issues to be watched for are also highlighted.

1 Introduction

Biological macromolecules tend to have a high degree of flexibility and under physiological conditions, they are usually found within a liquid solution. In such an environment, the macromolecules undergo various types of motion continuously. Excellent progress has been made in crystallography to capture the atomistically detailed structures of molecules in the solid state. However, crystallographic structures may not accurately resemble those that are found in solution, nor do they provide information about the dynamics of the molecule under realistic physiological conditions. Since the molecules are moving in solution, it is difficult to obtain an atomistically detailed 'snapshot' experimentally. Instead, one can obtain orientationally averaged scattering data of the macromolecules in solution and then use computationally generated, physically plausible models of the macromolecule in order to deduce the structure and dynamics of the actual molecules.

Many different techniques are available for simulating the dynamics involving soft materials and the particular choice depends on the level of detail required and the amount of computational power at the user's disposal. For instance, quantum mechanical methods, such as density functional theory, consider details down to the level of electron clouds within the individual atoms making up a molecule. This makes quantum mechanical techniques computationally demanding and very slow for large molecules. In macromolecules, where such precision is rarely required, techniques based on classical laws of physics are better suited in order to ensure that a sufficiently large range of molecular configurations are explored within a manageable computational time. This introduces the concept of *coarse*

* Corresponding author: bhatt@physics.org

graining in simulations. As the smallest unit of the material being simulated gets larger, the system is referred to as coarse-grained.

This chapter aims to provide a simplified introduction for biologists to two widely used techniques for computationally simulating the structure and dynamics of biological macromolecules under realistic interactions and constraints. In Sections 2 and 3 we go through the theoretical background behind the techniques of molecular dynamics (MD) and Monte Carlo (MC) simulations. After enlisting some of the issues and pitfalls that may arise in the simulations and possible ways to avoid them in Section 4, a typical workflow involved in using small angle neutron scattering (SANS) data to deduce solution structure of biological macromolecules is discussed in Section 5.

2 Molecular dynamics basics

When two atoms are placed close to each other, such as is the case in a molecule, they would, in general, exert mutual forces on each other. This is essentially due to the fact that atoms are not solid spheres with hard boundaries, but are enveloped by rather fuzzy probability clouds of negatively charged electrons that fade away gradually as one moves away from the centre of an atom. If the value of this mutual force, **F**, is known, one can solve the two simultaneous equations, one for each atom, resulting from Newton's second law of motion:

$$\begin{aligned} \mathbf{F} &= m_1\mathbf{a}_1 \\ -\mathbf{F} &= m_2\mathbf{a}_2, \end{aligned} \tag{1}$$

where m_1 and m_2 are the masses of the two atoms and $\mathbf{a}_1$ and $\mathbf{a}_2$ are the accelerations of the two atoms that result from this equal and opposite force acting on each other. Since an exact mathematical solution for $\mathbf{a}_1$ and $\mathbf{a}_2$ can be obtained in this way, the system is known as *analytically solvable* and we can recreate and predict the positions of the two atoms in three-dimensional space as a function of time. If, however, one adds just one more atom to the system, it turns out that the resultant three simultaneous equations of motion cannot be solved analytically. Indeed, for a molecule with a large number of atoms, this becomes an *N*-body problem, which has no analytical solution and must be solved numerically with the help of a computer.

In a classical *molecular dynamics* simulation, trajectories of atoms and molecules are obtained in real time and space by numerically solving the appropriate (and usually large) set of time-dependent equations of motion. It is known as a *deterministic* technique, since the state of the system at a given time step can be *predicted* using the knowledge of the state of the system at the previous time step, provided that the time step is adequately small and that information about the manner in which the atoms (or ions) interact with each other is available. MD is widely used to understand the structure, dynamics, phase transformation and surface phenomena in soft matter and hard matter systems. MD simulations that incorporate the principles of quantum mechanics, instead of Newtonian mechanics, also exist; however, for macromolecules these are prohibitively expensive in terms of their execution time with the current generation of computers. Hence, we focus our discussion on classical MD.

2.1 Trajectory calculation

Assuming that the total force $\mathbf{F}_j$ acting on each of the atoms $j = \{1, \ldots, N\}$, where N is the number of atoms in the system, is known, a common numerical scheme to obtain the positions of all the atoms at successive time steps is the Verlet algorithm [1]. It is derived by Taylor expanding the position $\mathbf{r}_i(t)$ up to third order in both forward and reverse time and then adding the resulting two expressions. The end result is

$$\mathbf{r}_i(t + \Delta t) = 2\mathbf{r}_i(t) - \mathbf{r}_i(t - \Delta t) + \frac{\mathbf{F}_i(t)}{m_i}(\Delta t)^2 \tag{2}$$

which allows for the atomic positions to be calculated at time $t + \Delta t$ given the positions at times t and $t - \Delta t$. In other words, it allows one to numerically propagate the trajectory forward in time using an existing set of atomic coordinates and forces. Other more sophisticated methods, such as the leapfrog algorithm, also exist [2].

The utility of Eq. (2) is conditional upon the time step Δt being adequately small. As a rule of thumb, it should be smaller than the timescale of the fastest mode of movement present in the given system, e.g. the thermal vibrations of the atoms making up a molecule, in order to be able to capture the dynamics. Under room temperature and pressure conditions, atoms in a molecule typically vibrate at frequencies of $10^{13} - 10^{14}$ Hz, which means a time step of a few femtoseconds (10^{-15} s) would work without the simulation crashing. A time step smaller that this would slow down the simulation unnecessarily without much gain in the accuracy of the results.

2.2 Force fields

The force **F** appearing in Eq. (1) is derived from the interatomic interactions that are acting among the atoms (and ions) in the molecule. Naturally, defining these interactions accurately is the most crucial part of an MD simulation, as the level of accuracy in these dictates the trustworthiness of the evolution and the emergent structure of the molecule. Let us therefore take a qualitative overview of what constitutes a typical set of interatomic interactions, also known as the force field, that go into an MD simulation.

An interaction between two atoms labelled i and j is defined in terms of the interatomic potential energy, $U(r_{ij})$, that exists between these two atoms solely. The force experienced by atom j, $\mathbf{F}_j$, as a result of atom i is related to the pair-potential energy $U(r_{ij})$ through the relation

$$\mathbf{F}_j = -\frac{1}{r_{ij}}\left[\frac{\partial U(r_{ij})}{\partial r_{ij}}\right]\mathbf{r}_{ij}, \tag{3}$$

where $\mathbf{r}_i$ and $\mathbf{r}_j$ are the Cartesian coordinates of the two atoms with respect to an arbitrarily defined origin in the simulation cell, while

$$\mathbf{r}_{ij} = \mathbf{r}_j - \mathbf{r}_i \tag{4}$$

is the vector from the jth atom to the ith atom and $r_{ij} = |\mathbf{r}_{ij}|$ is the scalar magnitude of this vector. The function $U(r_{ij})$ is, somewhat confusingly, often referred to as simply *potential* in the MD community. For the purpose of tractability, it is usually split into different parts that describe the various components of the overall interaction.

As a first split, the total potential energy, U_{Total}, is expressed as a sum of the potential energy due to chemical bonds, U_{Bonded}, and the potential energy between chemically non-bonded atoms, $U_{\text{Non-bonded}}$:

$$U_{\text{Total}} = U_{\text{Bonded}} + U_{\text{Non-bonded}}. \tag{5}$$

The term U_{Bonded} can be further split as

$$U_{\mathrm{Bonded}} = U_{\mathrm{Stretching}} + U_{\mathrm{Bending}} + U_{\mathrm{Torsion}} + U_{\mathrm{Improper}}, \tag{6}$$

where $U_{\mathrm{Stretching}}$ represents the stretching motion between two chemically bonded atoms, U_{Bending} describes the bending of the bonds among three atoms, U_{Torsion} is the dihedral angle potential to allow for the twisting movement among atoms and U_{Improper} represents dihedral potentials that are often used to restrict the geometry of molecules (e.g., to maintain the planar geometry of a molecule). Each of these four terms are harmonic in nature (i.e. an oscillating pattern around a mean central position). However, one could impose a rigid bond if required by replacing a given term with a constraint.

Non-bonded interactions are expressed as a sum of the electrostatic and van der Waals interactions:

$$U_{\mathrm{Non\text{-}bonded}} = U_{\mathrm{Electrostatic}} + U_{\mathrm{van\text{-}der\text{-}Waals}}. \tag{7}$$

Here, $U_{\mathrm{Electrostatic}}$ is the energy due to the Coulombic interaction between two electrical charges, while $U_{\mathrm{van\text{-}der\text{-}Waals}}$ takes into account dipole formation and the overlap of the electron clouds of two (or more) atoms. The van der Waals component of the interaction is very important, since it ensures that two ions with opposite charges placed close to each other do not end up sitting on top of each other, as would happen if only the electrostatic interaction were to be present. A considerable amount of effort usually goes into determining the exact form of this part of the interaction as it also dictates the equilibrium distance between two atoms (or ions).

One popular form that is chosen to represent the van der Waals interaction is the Lennard-Jones (L-J) potential, also sometimes referred to as 6-12 or 12-6 potential. Its functional form is:

$$U_{\mathrm{L\text{-}J}}(r_{ij}) = 4\epsilon\left[\left(\frac{\sigma}{r_{ij}}\right)^{12} - \left(\frac{\sigma}{r_{ij}}\right)^{6}\right] \tag{8}$$

To understand this potential, it is useful to sketch the magnitude of $U_{\mathrm{L\text{-}J}}$ as a function of the interatomic distance, r_{ij}, as shown in Figure 1. The parameter ϵ is the 'depth' of the L-J potential energy and is equal to the minimum energy required to break the van der Waals bond between atoms i and j, while σ is the interatomic finite distance at which the L-J potential energy is zero. Since the interatomic force is defined in terms of the slope of the potential energy, the minimum of the curve in Figure 1, marked by a red dot, represents the equilibrium point where the slope, and hence the interatomic force, is zero. The part of the curve to the left of this point represents repulsive force between the atoms, which in the real world occurs due to the overlap of the electron clouds of the atoms, while the part to the right characterises attractive force. Hence, this form of the potential is a convenient way of mimicking the spring-like character of the forces that exist between neutral atoms in the real world: when two atoms are pulled apart away from the equilibrium distance, they will tend to attract each other; however, if the two atoms come closer to each other than the equilibrium distance, they would experience a rapidly increasing repulsive force that will ensure that the two atoms do not quite merge into each other and will move away quickly. It also follows that at a non-zero finite temperature, two atoms placed close to each other will thus tend to perform oscillatory motion about the equilibrium point.

To obtain the value of the equilibrium distance r_{eq}, one can evaluate the first derivative of Eq. (8) with respect to r_{ij} and equate it to zero. This leads to

$$r_{\text{eq}} = 2^{1/6}\sigma \tag{9}$$

Equation (8) can therefore be rewritten as

$$U_{\text{L-J}}(r_{ij}) = \epsilon\left[\left(\frac{r_{\text{eq}}}{r_{ij}}\right)^{12} - 2\left(\frac{r_{\text{eq}}}{r_{ij}}\right)^{6}\right]. \tag{10}$$

In literature, both these forms of $U_{\text{L-J}}(r_{ij})$ can be found since both of them are equivalent. In a typical MD software, only the numerical values of ϵ, in addition to σ or r_{eq}, depending on whether Eq. (8) or (10) is implemented in the software, need to be provided while defining the L-J potential.

Another parameter that requires to be specified in an MD simulation is the cut-off distance, r_{cut}. This is the distance from a given atom when both the potential energy and the interatomic force are assumed to have fallen to negligible values. In Figure 1 it can be seen that at $r_{ij} = r_{\text{cut}}$ not only $U_{\text{L-J}}(r_{ij})$ is very close to zero, the slope of the curve is also very small, making the force negligible according to Eq. (3). Hence, considering interactions beyond this radius of influence would be a waste of computational effort. It is therefore necessary to choose this distance wisely. A large value of r_{cut} may slow down the simulation unnecessarily without perceptible gain in the accuracy of the results, while choosing too small a value would risk sacrificing the accuracy. For biomolecular simulations, $r_{\text{cut}} \approx 15$ Å is usually adequate, although values as small as 10 Å are not uncommon in literature.

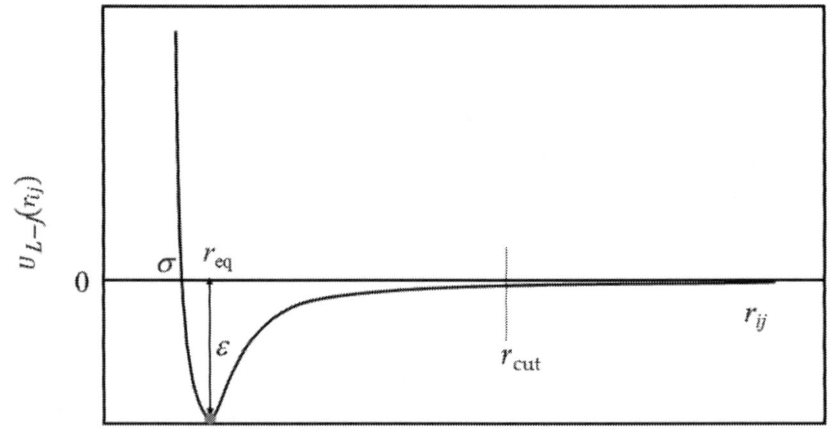

Fig. 1: Lennard-Jones interaction curve between two non-bonded atoms. The red dot marks the point of equilibrium between the two atoms.

All the rest of the terms on the right-hand side of equations (5), (6) and (7) may also have a mathematical expression of their own – often quite complex – or some of these can be expressed through a lookup table of numerical values as a function of distance, for example. A full discussion of these is beyond the scope of this chapter; however, for biomolecules well established parameter files built up by experts through many years of research, often using quantum mechanical considerations, are usually available for download. Some of the well-known biomolecular potentials include AMBER [3], BMS [4], CHARMM [5], ECEPP [6], GROMOS [7-9], MM3 [10], OPLS [11] and PFF [12], although this list is by no means exhaustive.

Some of the popular, freely available MD packages include DL_POLY [13], NAMD [14], GROMACS [15], CHARMM [16] and LAMMPS [17]. Each of these packages have their own appeal and individuals' choice is often based on the exact nature of the system to be simulated.

3 Monte Carlo approach

While MD is a deterministic technique, whereby the positions of the atoms evolve in a sequential manner from one time step to another following the laws of mechanics, Monte Carlo is known as a *stochastic* or *non-deterministic* technique. In MC, the state of the system from one step to another evolves in a random manner, albeit within physical constraints imposed by factors such as chemical bonds, force fields, temperature, pressure and so forth. Not all of these constraints need to be imposed in a particular simulation and this is something to be decided by the software designers and the users.

In order to understand how a molecular system can be probed through random means, one needs to first consider the concept of *potential energy surface* in physics, which, contrary to the common meaning of the word 'surface', could in general be a n-dimensional entity, where n is the number of degrees of freedom present in the system. Figure 1 is an example of an energy curve with only one degree of freedom, namely the distance between two atoms. In Figure 2, an energy surface with two degrees of freedom, denoted as R_1 and R_2, is shown (it would be impossible to graphically depict an energy surface with three or more degrees of freedom). It consists of hills, valleys and saddle points. Each minimum within a given valley is known as a *local minimum*, while the absolute lowest point within the entire surface is known as the *global minimum*. This kind of shape of the energy surface is generally caused as a result of the competing forces that an atom experiences from different directions. Since the atoms are constantly trying to minimise their own energy, one would expect the atoms to spend most of the time in regions around the local minima. Excursion onto higher energy points occurs as a result of temperature, which drives thermal fluctuations, which occasionally also leads to an atom climbing all the way up a hill and dropping into a neighbouring local minimum. Clearly then, the higher the energy, the lower the probability of finding atoms in that energy state and vice versa.

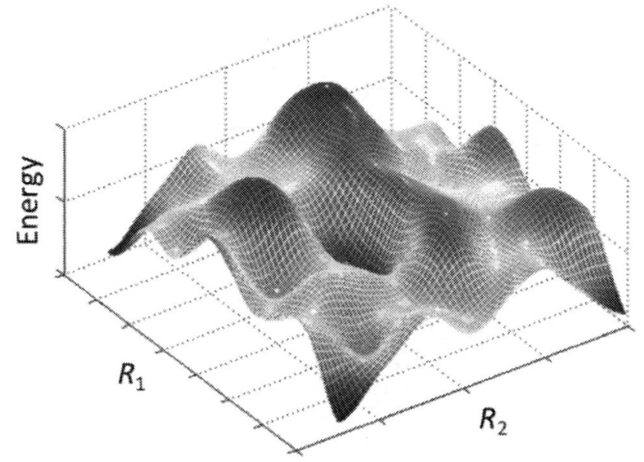

Fig. 2: Typical energy surface of a system with two degrees of freedom, labelled as R_1 and R_2.

It is this key observation that is utilised cleverly in a Monte Carlo simulation involving molecular structure. If **q** and **p** are two large vectors specifying the instantaneous positions and momenta, respectively, of all the atoms in the system, i.e.

$$\mathbf{q} = (x_1, y_1, z_1, x_2, y_2, z_2, \ldots, x_N, y_N, z_N) \tag{11}$$

$$\mathbf{p} = (p_{x,1}, p_{y,1}, p_{z,1}, p_{x,2}, p_{y,2}, p_{z,2}, \ldots, p_{x,N}, p_{y,N}, p_{z,N}), \tag{12}$$

then the space containing all the possible combinations of $(\mathbf{q}, \mathbf{p})$ is known as the *phase space* of the given system. The phase space contains all the information that is necessary to describe the state of the system. The probability $P(\mathbf{q}, \mathbf{p})$ of finding the system at a particular phase point is connected with the system energy, $E(\mathbf{q}, \mathbf{p})$ through the equation

$$P(\mathbf{q}, \mathbf{p}) = Z^{-1} \exp\left[\frac{-E(\mathbf{q}, \mathbf{p})}{k_B T}\right], \tag{13}$$

where k_B is Boltzmann's constant and T is the system temperature. The parameter Z is the system partition function given by

$$Z = \iint \exp\left[\frac{-E(\mathbf{q}, \mathbf{p})}{k_B T}\right] \mathrm{d}\mathbf{q}\mathrm{d}\mathbf{p}, \tag{14}$$

which acts to normalise $P(\mathbf{q}, \mathbf{p})$. Equation (13) provides a link between the probability, which is a central ingredient of the Monte Carlo philosophy, and the energy, which can be calculated using the position and momenta of a molecular system. This allows one to generate random configurations of the molecular system within the confines of the laws of physics, thus leading to physically realistic configurations.

To achieve this, Metropolis *et al.* proposed an algorithm [18], which was subsequently expanded by Hastings [19]. Figure 3 shows a flowchart of the Metropolis sampling algorithm for molecular structural studies. After setting up the initial coordinates of the atoms, an atom is picked randomly and attempt is made to move it to a new random position. If the energy of the new position is lower than that of the starting position, the new position is accepted because molecules always prefer lowering their energy; hence this is a perfectly plausible move. On the other hand, if the new trial position is at a higher energy, one does not reject the position outrightly and gives it a chance to succeed: the probability ratio,

$$p = \frac{P \text{ at new position}}{P \text{ at old position}} = \frac{\exp(-E_{\text{new}}/k_B T)}{\exp(-E_{\text{old}}/k_B T)} = \exp\left(-\frac{\Delta E}{k_B T}\right), \tag{15}$$

where $\Delta E = E_{\text{new}} - E_{\text{old}}$, is calculated and is compared to a random number between 0 and 1. If the random number is between 0 and p, the move is accepted; if the random number is greater than p, the move is rejected [20]. This strategy ensures that a Boltzmann distribution of molecular configurations is produced in accordance with Eq. (13). Far more configurations are produced around the valleys of the energy surface and only a few are produced around the peaks, which means a considerable amount of computational effort is saved that would otherwise be wasted in probing the higher energy configurations that are physically less plausible.

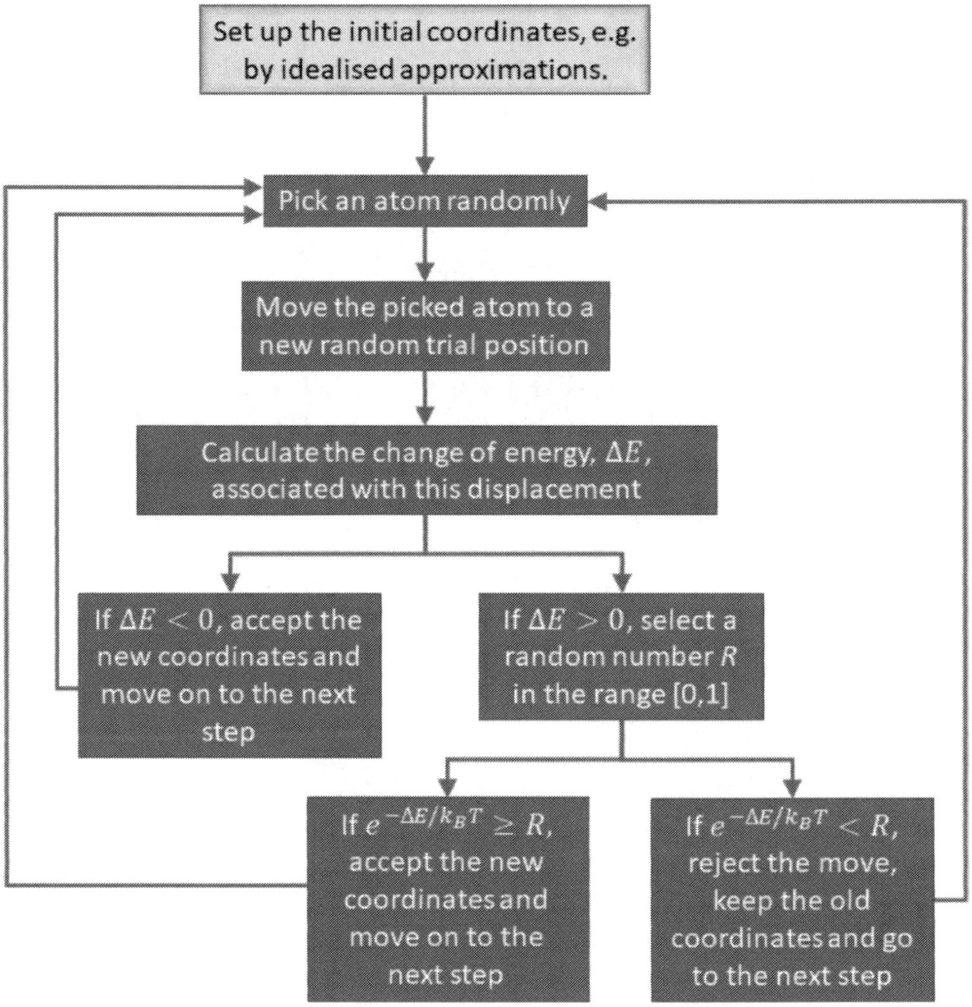

Fig. 3: Metropolis sampling algorithm for the structural studies of molecular systems.

Once the physically plausible configurations are produced, the average of a property B that is dependent on the atomic positions can be calculated using the simple relation

$$\langle B \rangle = \frac{1}{M} \sum_{i=1}^{M} B(\mathbf{q}_i), \tag{16}$$

where M is the number of accepted configurations.

Since there are no equations of motion to be solved in Monte Carlo, it is much faster than molecular dynamics in terms of probing points in the phase space that are truly useful. Also note that, unlike an MD simulation, there is no time coordinate in the MC recipe described above; the system jumps from one state to another random state so as to ensure that the best statistical ensemble of energetically plausible states is collected. If required, there is another class of MC simulations, known as *kinetic Monte Carlo* (kMC), which does allow an explicit time coordinate. This is made possible by introducing rates of various events, which are inversely proportional to time. An event is essentially seen as a transition between one state to another and the rate of this is connected to the energy barrier that exists between the two states. We shall not discuss kMC in detail here since time dependence is not required in the structural studies of macromolecules.

4 Practical considerations

In this section we enlist a few practical points to be considered during a typical workflow of molecular simulation.

4.1 Connectivity in molecules

Specifying only the atomic coordinates, for instance through a PDB file, is not sufficient to carry out a molecular simulation. One also needs to provide information about the connectivity of atoms within a molecule. Many simulation packages have their own ways of defining the different types of chemical bonds that exist between the atoms. For biomolecules, one 'standard' way of doing this is through a PSF file (Protein Structure File). Despite what the name may suggest, PSF files are used to define connectivity within non-protein biomolecules as well. Automated ways of creating a PSF file from a given PDB file include the online tool CHARMM-GUI [21] and the AutoPSF plugin included in the VMD package [22]. CHARMM-GUI also has a tool to check for problems in a given PDB file and to some degree fix the file such that it is compatible with the CHARMM force field. It also has a tool to detect glycans and perform any necessary modifications in the PDB file arising from this. In case one of the automatic methods fails for some reason, a semi-automatic tool to create the PSF file is the psfgen tool that comes with VMD. It is semi-automatic in that a set of topology files and an input file specifying the type of bonds that are present among the various atomic species need to be provided by the user. While this may require careful treatment, it does give the user more precise control. The AutoPSF plugin in VMD, in fact, implements psfgen under the hood. VMD is also a powerful tool to visualise and manipulate PDB files and trajectory files that are output by many simulation packages.

4.2 Speed-up through coarse graining

Biomolecules can be very large for a molecular simulation. A typical protein may contain tens of thousands of atoms, while a DNA molecule may contain several billions of atoms. Obtaining explicit variation among all these atoms is not only computationally expensive, but also unnecessary in a large number of cases. To speed up the simulation, one may perform selective coarse graining by defining large regions of the molecule as 'rigid' and only small parts of the molecule as 'flexible'. The rules of structural variation, according to either MD or MC, are then applied explicitly to all the atoms in the flexible regions and only to a representative point in each rigid region, which is usually the mid-point. The rigid regions can be identified by looking at the initial structure of the molecule. For instance, in a protein molecule one would find β-sheets that are made of several β-strands connected together and where motion beyond thermal vibrations does not occur. Assuming that the β-sheets are stable under normal physiological conditions, one could define these as rigid. Visibly thin regions of the molecule, such as linkers between domains, can be defined as flexible in order to drive structural variation. However, care must be taken so as to ensure that important bonds such as disulphide bridges do not appear within a flexible region in a Monte Carlo simulation, as sudden variation from one state to another may break these bonds.

5 Inference of solution structure from scattering data

Once an ensemble of structures has been created by a simulation, calculating small angle neutron scattering (SANS) profiles is conceptually straightforward, although it may require some processing effort. Computationally one performs essentially the same procedure as in a real neutron scattering experiment, but this time in a virtual world, with a large set of PDB files of the molecular structures created by a simulation replacing the molecules that exist in a real experiment. A typical workflow of the whole process is shown in Figure 4.

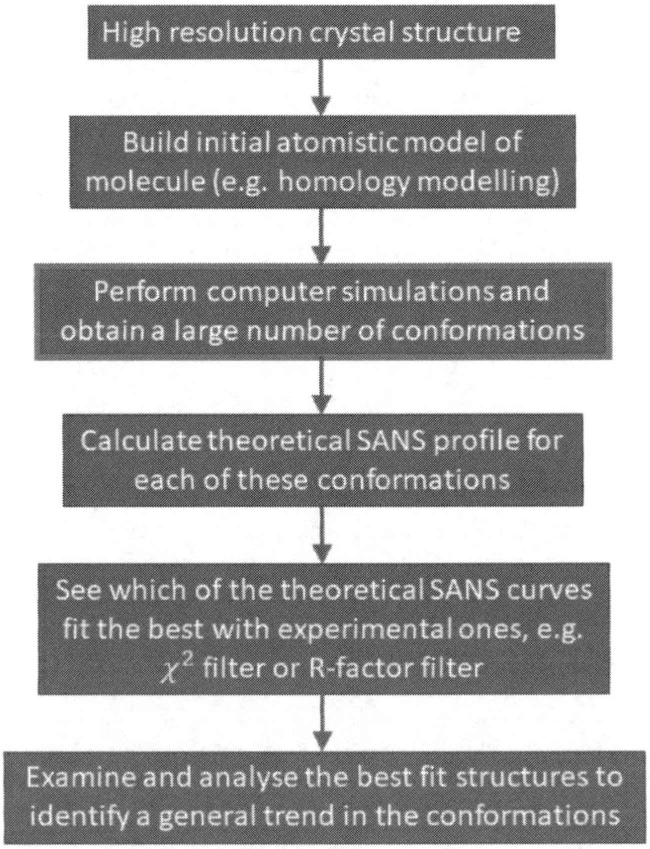

Fig. 4: A typical workflow to infer atomistic structures of molecules in solution using SANS data.

5.1 Theoretical scattering profiles

The step marked with a red box is already described in Sections 2 and 3. The next step is to calculate theoretical SANS profile for each of the conformations produced by the simulations. One way of doing this is through the Debye formula for spherically averaged one-dimensional scattering intensity [23],

$$I(Q) = \sum_{i=1}^{M}\sum_{j=1}^{M} f_i(Q)f_j(Q)\frac{\sin(Q \cdot r_{ij})}{Q \cdot r_{ij}}, \quad (17)$$

where Q is the scattering vector, $I(Q)$ is the scattered intensity of the neutrons, r_{ij} is the distance between atoms labelled i and j within the given theoretical conformation, $f_i(Q)$ is the form factor (or scattering factor) for atom i, which describes the scattering contribution of the stand-alone atom, and M is the number of atoms in the molecule. Provided that the form factor is known for all the atomic species found in a biomolecule, only the interatomic distances, which are available from the generated PDB files, are required to calculate the scattering profiles from Eq. (17).

Implementation of Eq. (17) requires calculating M^2 terms for each conformation, which can be computationally demanding when the molecule is large (large M) and there are a large number of conformations to go through. Approximation techniques have been reported that aim to reduce the burden of these calculations. These include the multipole expansion (ME) technique of Svergun *et al.* [24] and the golden vector method of Watson and Curtis [25] that scales linearly with the number of atoms in the molecule.

One thing that must be taken care of while calculating the theoretical scattering profiles is that the Q-values must match those that are present in the experimentally obtained scattering curves. This is so that $I(Q)$ for both the theoretical and experimental curves can be compared at exactly the same set of Q-values. If this is not possible to achieve, one can interpolate the experimental data points at the desired set of Q-values using, for instance, a spline method. If there is a mismatch in the Q-values, there are likely to be significant errors in the subsequent analysis of the results; hence, this is an important step.

5.2 Identification of best-fit structures

The quality of a computationally generated atomistic conformation of the molecule can be tested by performing a statistical hypothesis test, such as an R-factor test (or sometime a χ^2 test). The R-factor is defined as

$$R\text{-factor as percentage} = \frac{\sum_i \left||I_{\text{Expt}}(Q_i)| - \eta |I_{\text{Theor}}(Q_i)|\right|}{\sum_i |I_{\text{Expt}}(Q_i)|} \times 100, \quad (18)$$

where $I_{\text{Expt}}(Q_i)$ and $I_{\text{Theor}}(Q_i)$ denote the experimentally obtained and theoretically calculated scattering intensities, η is a scaling factor used to match the theoretical curve to the experimental $I(0)$ value and i runs over the range of all the Q-values over which the comparison of the curves is to be made. This method compares the scattering profile of each of the theoretical conformations and produces a numerical value of R-factor that indicates how closely the curve matches with the target experimental curve (Figure 5). The lower the R-factor, the better the fit.

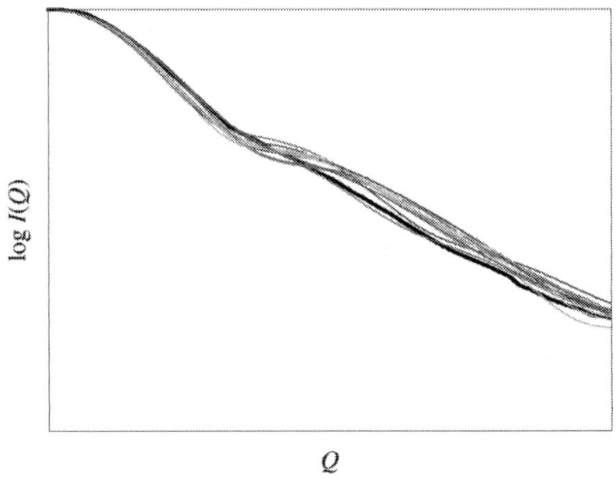

Fig. 5: Comparison of the theoretically produced scattering profiles (non-black colours) with a single experimental profile (black).

If one calculates the R-factor for all the computationally generated molecular conformations as a function of the radius of gyration, R_g, of each conformation, a plot similar to the one shown in Figure 6 should come out, provided that a sufficiently large number of physically plausible molecular conformations have been used to produce such a plot. The plot should have a U-shape with a distinct minimum surrounded by rising values of R-factor on either side of the minimum. This allows one to assert with confidence that the conformations with the lowest R-factors are indeed the best-fit structures among the computationally generated ensemble. In any case, one should aim for R-factors of below 5% as a sign of a reasonably good fit. Hence, if the plot never dips below 5%, then either insufficient number of conformations were generated, or their quality was questionable, or perhaps the quality of the experimental scattering curve itself needs to be investigated.

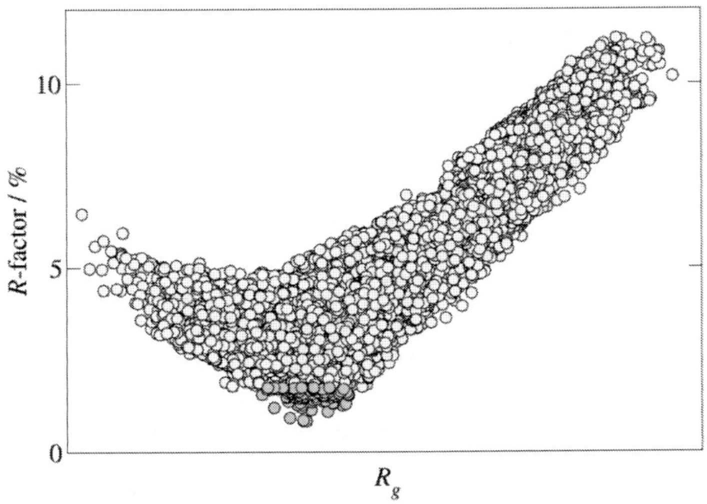

Fig. 6: Identification of the best-fit structures using the R-factor filter. Each data point (circles) in this plot corresponds to a unique atomistic conformation produced by the molecular simulation. Yellow circles represent all the conformations that were created, while the best-fit 100 conformations with the lowest R-factors are shown in green.

A certain number of best-fit molecular structures identified in this way (e.g. the green points in Figure 6) can be subjected to further statistical or numerical analysis in order to reveal a common pattern within them. It must be stressed that a single structure is *not* sought to be identified here. Rather, a tightly-defined ensemble of structures is being obtained to match the solution structure of the biomolecules that are giving rise to the spherically averaged scattering curve in SANS experiments.

5.3 Web-based workflow

All the above steps outlined in this Chapter can be fairly demanding task for a non-expert. Preparing an input PDB file for simulation and then preparing the (complex) additional input files for performing the molecular simulations are not trivial tasks even for experts in the field. Not to mention the code to be written for the subsequent analysis of the results. An online free-to-use portal, SASSIE-web, simplifies this task considerably by presenting highly abridged web-based input forms for each of the tasks in the workflow [26]. It has modules to prepare structures, carry out simulations, calculate theoretical SAXS and SANS data and compare the outcome to experimental data. While the key outputs are shown on the web page after the calculations, one can download the more detailed files for further analysis.

6 Conclusions

Deducing the solution structure of biological macromolecules in solution is a complex task that requires expertise of several fields to be applied together. While SANS experiments are an effective way of capturing spherically averaged scattering profiles of the macromolecules, a computational counterpart of this process is required to exploit and decode the information provided by the experimental data. To this end, structural variation in a macromolecule can be performed using molecular simulations. Theoretical scattering profiles calculated from the computationally generated structures can be compared with the experimental SANS curve to obtain a set of conformations that best reflect the structure of the macromolecule in solution.

References

1. L. Verlet, Phys. Rev. **159**, 98 (1967).
2. M. P. Allen, D. J. Tildesley, *Computer Simulation of Liquids* (Clarendon Press, Oxford, 1989).
3. S. J. Weiner, P. A. Kollman, D. T. Nguyen, D. A. Case, J. Comp. Chem. **7**, 230 (1986).
4. D. R. Langley, J. Biomol. Struct. Dyn. **16**, 487 (1998).
5. K. Vanommeslaeghe, E. Hatcher, C. Acharya, S. Kundu, S. Zhong, J. Shim, E. Darian, O. Guvench, P. Lopes, I. Vorobyov, A. D. MacKerell jr., J. Comput. Chem. **31**, 671 (2010).
6. G. Nemethy, M. S. Pottle, H. Scheraga, J. Phys. Chem. **87**, 1883 (1983).
7. W. F. van Gunsteren and H. J. C. Berendsen, *Groningen Molecular Simulation (GROMOS) Library Manual* (BIOMOS, Nijenborgh, 9747 Ag Groningen, 1987).
8. C. Oostenbrink, A. Villa, A. E. Mark, W. J. van Gunsteren, Comput. Chem. **25**, 1656 (2004).
9. T. A. Soares *et al.*, J. Comput. Chem. **26**, 725 (2005).
10. N. L. Allinger, Y. H. Yuh, J.-H. Lii, J. Am. Chem. Soc. **111**, 8551 (1998).
11. W. L. Jorgensen, J. D. Madura, C. J. Swenson, J. Amer. Chem. Soc. **106**, 6638 (1984).
12. G. A. Kaminski *et al.*, J. Comput. Chem. **23**, 1515 (2002).
13. I. T. Todorov, W. Smith, K. Trachenko, M. T. Dove, J. Mater. Chem. **16**, 1911 (2006).
14. J. C. Phillips, R. Braun, W. Wang, J. Gumbart, E. Tajkhorshid, E. Villa, C. Chipot, R. D. Skeel, L. Kale, K. Schulten. J. Comput. Chem. **26**, 1802 (2005).
15. M. J. Abraham, T. Murtola, R. Schulz, S. Páll, J. C. Smith, B. Hess, E. Lindahl, SoftwareX **1-2**, 19 (2015).
16. B. R. Brooks *et al.*, J. Comput. Chem. **30**, 1545 (2009).
17. S. Plimpton, J. Comp. Phys. **117**, 1 (1995).
18. N. Metropolis, A. W. Rosenbluth, M. N. Rosenbluth, A.H. Teller, E. Teller, J. Chem. Phys. **21**, 1087 (1953).
19. W. K. Hastings, Biometrika **57**, 97 (1970).
20. C. J. Cramer, *Essentials of Computational Chemistry* (John Wiley & Sons, 2nd edn., 2004).
21. S. Jo, T. Kim, V. G. Iyer, and W. Im, J. Comput. Chem. **29**, 1859 (2008).
22. W. Humphrey, A. Dalke, K., J. Schulten, Molec. Graphics **14**, 33 (1996).
23. P. Debye, Ann. Phys. (Leipzig), **46**, 809 (1915).
24. D. Svergun, C. Barberato, M. H. J. Koch, J. Appl. Cryst. **28**, 768 (1995).
25. M. C. Watson and J. E. Curtis, J. Appl. Cryst. **46**, 1171 (2013).
26. S.J. Perkins *et al.*, J. Appl. Crystall. **49**, 1861 (2016).

EPJ Web of Conferences **236**, 04001 (2020) https://doi.org/10.1051/epjconf/202023604001
JDN 24

An introduction to neutron reflectometry

Fabrice Cousin[1,*] and *Giulia* Fadda[1,2]

[1]Université Paris-Saclay, Laboratoire Léon-Brillouin, CEA-CNRS, CEA-Saclay, F-91191 Gif-sur-Yvette, France
[2]Université Sorbonne Paris Nord, UFR SMBH, 74 rue Marcel-Cachin, F-93017 Bobigny, France

Abstract. Specular neutron reflectivity is a neutron diffraction technique that provides information about the structure of surfaces or thin films. It enables the measurement of the neutron scattering length density profile perpendicular to the plane of a surface or an interface, and thereby gives access to the profile of the chemical composition of the film. The wave-particle duality allows to describe neutrons as waves; at an interface between two media of different refractive indexes, neutrons are partially reflected and refracted by the interface. Interferences can occur between waves reflected at the top and at the bottom of a thin film at an interface, which gives rise to interference fringes in the reflectivity profile directly related to its thickness. The characteristic sizes that can be probed range from 5Å to 2000 Å. Neutron-matter interaction directly occurs between neutron and the atom nuclei, which enable to tune the contrast by isotopic substitution. This makes it particularly interesting in the fields of soft matter and biophysics. This course is composed of two parts describing respectively its principle and the experimental aspects of the method (instruments, samples). Examples of applications of neutron reflectometry in the biological domain are presented by Y. Gerelli in the book section "Applications of neutron reflectometry in biology".

1 Introduction

Neutron reflectometry (NR) is a surface technique that enables to measure the thickness and the chemical composition of one or several thin layers at a surface or an interface. Typical neutron wavelengths used during experiments range from 3Å to 25Å, which allows to study thicknesses from 5Å to 2000Å. A neutron beam is sent, at a grazing angle, to the interface or the surface under study in order to measure the reflectivity coefficient R of the neutron beam. In this course, we will only focus on the specular reflectivity that corresponds to the case where the incident and reflected neutron beams are symmetric with respect to the surface normal. This provides the profile of chemical composition perpendicular to the surface, averaged on the whole surface. The principle of off-specular studies that treat the case of neutrons that are scattered out of the specular direction and allow to access to in-plane structural information can be found elsewhere [1].

The lower limit of thickness that can be measured is directly linked to the minimal reflectivity R that can be experimentally obtained ($R \sim 10^{-6}$ to 10^{-7} on samples of a few cm^2) with the current brightness of neutrons sources. Such a limit can usually be extended to lower values by X-ray reflectivity, the counterpart experiment of NR with photons of the same wavelengths, and which has thus to be preferred to NR. Nevertheless, NR offers unique

* Corresponding author: fabrice.cousin@cea.fr

advantages that make it irreplaceable in many experimental cases: neutrons can access to buried interfaces, they have a high sensibility for materials containing hydrogen like biological systems and they can probe magnetic properties of materials at the atomic level. This is due to some features of the neutron-matter interaction, as detailed in the following.

First, neutron interacts directly with atom nuclei. The interaction has a very short range and is described by the coherent scattering length b. It varies strongly between isotopes of the same chemical element and randomly across the elements of the periodic table, and can be, either positive, or negative. In particular, the b value for hydrogen ^1H is negative and has a very different value with respect to all other atoms constitutive of organic molecules (C, O, N, ...) and importantly of deuterium ^2H. This enables to achieve experiments on organic materials because the neutron refraction index of a given organic molecule is essentially dependent on its content on hydrogen atoms. Such refraction index can be varied on a large range by replacing some hydrogen atoms by deuterium atoms. Such replacement does not alter most of the physical properties of the molecules, but may change slightly some of them, *eg* hydrogen bonding. This opens the way to contrast variation experiments where some molecules are labelled with deuterium to create a neutron contrast in the system. It is also possible to continuously tune the neutron refractive index of a solvent in a complex system by mixing hydrogenated and deuterated solvents in order to match one component, making it invisible from the neutron point of view. Such contrast variation experiments have been very successfully applied in soft matter in the field of surfactants and polymers [2-4] or in biophysics [5-7].

Second, neutrons are very weakly absorbed by matter, except for few atoms of the periodic table that have a huge neutron absorption cross-section (for instance boron, gadolinium, or lithium). This weak interaction makes possible the study of buried interfaces or to achieve *in situ* experiments in various sample environments.

Third, NR experiments are not destructive, contrarily to X-ray experiments, because the energy of neutrons used for experiments is of a few meV.

It is finally worth noting that NR allows the measurement of the magnetization profile in thin magnetic films because neutrons bear a spin $\pm \frac{1}{2}$ and interact with the atoms having a magnetic moment; but this last aspect is out of the scope of this course focused on biology and is documented elsewhere [8].

Such specificities of neutrons have made NR experiments very popular since the early 90's. But it is important to keep in mind that it is advantageous to combine NR with other surface techniques like X-ray reflectivity, ellipsometry, and with techniques enabling to determine in-plane structure information like either scattering methods (GISAS, surface diffraction, off-specular measurements), or microcopy techniques (AFM, Brewster microscopy).

This course is divided into two parts. In the first one, the principle of NR will be presented. In the second part we will describe how to realize a NR experiment from a practical point of view (description of instruments and of possible geometries of measurements, sample environments, accessibility to neutron beams, neutron facilities…). The treatment and analysis of NR data, as well as examples of application of NR to biological systems, are described in the book section written by Y. Gerelli.

2 Principle of Neutron Reflectivity

The formalism described in this part is a wave formalism very close to the one usually used for reflection of light. It is limited to specular reflectivity and does not consider the absorption of neutrons.

2.1 Neutron-matter interaction and calculation of refractive index for neutrons

Let us consider a neutron beam reflecting at an infinite flat surface with an incident angle θ with respect to the plane z = 0, as showed in Figure 1. This surface is defined by the interface between the vacuum ($n = 1$) and a medium of refraction index n. In vacuum, the modulus of the wave vector of an incident neutron of wavelength λ is $k = 2\pi/\lambda$. In the medium of refractive index n, the neutron interacts with the nuclei of atoms. The amplitude of the interaction between a neutron and an atomic nucleus is given by the scattering length b, which has the dimension of a length; it may be regarded as a constant for a given isotope. Some typical values of b are given in Table 1. The neutron-matter interaction is described by the Fermi pseudo-potential [9]; it can be shown that the neutron wave vector k_n in a medium of refractive index n can be written:

$$k_n^2 = k^2 - 4\pi Nb \qquad (1)$$

where N is the number of atoms per volume unit. This last equation allows to express the refractive index n of a homogeneous medium, defined as the ratio of the wave vectors in the material and in vacuum:

$$n^2 = \frac{k_n^2}{k^2} = 1 - \frac{\lambda^2}{\pi} Nb \qquad (2)$$

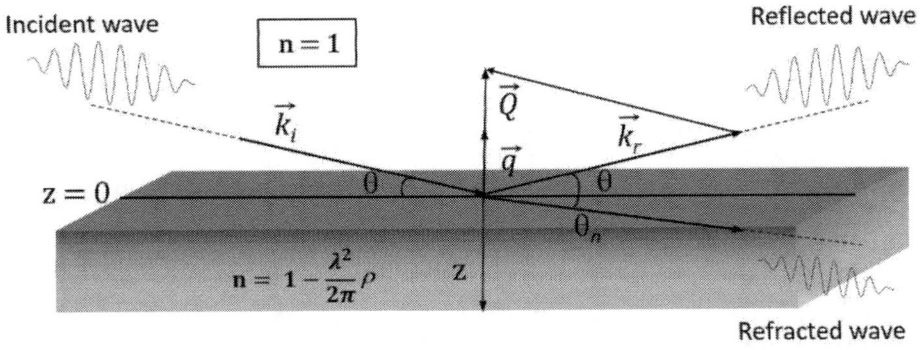

Fig. 1. Reflection on an infinite planar surface.

The product Nb is the **neutron coherent scattering length density (SLD)** ρ. Table 1 shows the coherent scattering length b of some atoms and the coherent scattering length density ρ of some materials [10]. Some full neutron tables of elements can be easily found on internet [11]. The SLD of the substrate is a crucial information in a NR experiment. It has to be ideally calculated prior to the experiment. Its calculation is very simple for crystals or solvents[†] but may raise difficulties for molecules in solution when their specific volumes are not known with high precision.

The characteristic wavelengths of the incident neutrons being always of the order of few Å and the SLD of the order of 10^{-6} Å^{-2}, the values of the neutron refractive index are always very close to 1 ($\sim 1 \pm 10^{-3}$-10^{-4}). Therefore, unlike the visible light, reflection/refraction for neutrons can only occur at very grazing angles, of the order of ~1° and the refractive index n for neutrons can be larger than 1.

Since n is very close to 1, a series expansion of eq. (2), enables to write:

$$\boxed{n \approx 1 - \frac{\lambda^2}{2\pi}\rho} \quad (3)$$

Table 1. Coherent scattering length of some atoms and scattering length density of some molecules. PC stands for phosphocholine.

Coherent scattering length b (10^{-12} cm)					
^1H	D (^2H)	C	O	N	Si
-0.374	0.667	0.665	0.580	0.936	0.415
Scattering Length Density ρ (10^{-6} Å^{-2})					
H_2O	D_2O	PC_H	PC_D	SiO_2	Si (crystal)
-0.56	6.38	-0.38	7.1	3.41	2.07

Analogously to the geometrical optics, refraction of neutrons at an interface between the vacuum and a medium of index n is described by the Snell-Descartes law:

$$\cos\theta = n\cos\theta_n \quad (4)$$

where the angles θ and θ_n are depicted in Figure 1. When $\theta_n = 0$, we can define the critical angle for total reflection θ_c:

$$\cos\theta_c = n \quad (5)$$

Neutrons are totally reflected for $\theta \leq \theta_c$. If n is higher than 1, the conditions of total reflection are never fulfilled, which happens for example at the air/water interface when the substrate is 100% H_2O. From equations (2) and (5), we obtain:

[†] Calculation of SLD in the case of water. Since the molar mass of H_2O is 18 g/mol and the density of H_2O is 1 g/cm^3, the volume of a molecule V of water is $18/6.02\ 10^{23} = 2.99\ 10^{-23}$ cm^3 *per* molecule, for both H_2O and D_2O. Since $\rho = (\Sigma_i\ b_i)/V$ where the b_i correspond to the atoms constitutive of the water molecule (table 1), one obtains:
$\rho(H_2O) = (-0.374*2+0.58).10^{-12} / 2.99\ 10^{-23} = -0.56\ 10^{10}$ cm^{-2} ($-$ 0.56 10^{-6} Å^{-2})
$\rho(D_2O) = (0.667*2+0.58).10^{-12} / 2.99\ 10^{-23} = 6.38\ 10^{10}$ cm^{-2} (6.38 10-6 Å^{-2})

$$sin\,\theta_c = \sqrt{\frac{\rho}{\pi}}\lambda \qquad (6)$$

When total reflection occurs, the measurement of the critical angle θ_c for a given wavelength is a very precise measurement of the substrate SLD and therefore of its chemical composition.

2.2 Reflection on a succession of layers

The reflectivity is a function that only depends on the scattering vector $\boldsymbol{Q}$ defined by:

$$\vec{Q} = \vec{k_r} - \vec{k_i}$$
$$Q = \left\|\vec{k_r} - \vec{k_i}\right\| = \frac{4\pi\,sin\,\theta}{\lambda} \qquad (7)$$

It is equal to twice the projection of the incident wave vector $\boldsymbol{k_r}$ on the z axis perpendicular to the surface, as depicted in Figure 1. In the following we will use in calculations the variable q, projection of the wave vector on z such that $\boldsymbol{q} = \boldsymbol{Q}/2$.

If the system under study is made of several layers, each of them having a refractive index n_p, neutrons will be partially reflected and transmitted from the layer p to the $p+1$ one. From equation (1) the scattering wave vector for each layer writes:

$$q_p^2 = q^2 - 4\pi\rho_p \qquad (8)$$
$$q_{p+1}^2 = q^2 - 4\pi\rho_{p+1} \qquad (8')$$

The continuity conditions for the wave functions and its derivative at the interface $(p/p+1)$ lead us to write the reflection coefficient R, which is defined as the ratio of the intensity of the beam reflected by the layer $p+1$ to the intensity of the incident beam on such a layer:

$$R = \frac{|B_p|^2}{|A_p|^2} = \left|\frac{q_p - q_{p+1}}{q_p + q_{p+1}}\right|^2 \qquad (9)$$

where A_p and B_p are respectively the amplitudes of the wave propagating towards the inner and the outer of the material.

Using this equation recursively it is possible to calculate the whole reflectivity of the system. This method is known as the optical matrix method and uses the Abélès formalism [12].

It is easy to show that reflectivity from a simple single interface between two media of refractive index n_0 and n_1, respectively, can also be written:

$$R = \left|\frac{n_0 sin\theta_0 - n_1 sin\theta_1}{n_0 sin\theta_0 + n_1 sin\theta_1}\right|^2 \qquad (10)$$

from Snell-Descartes condition we can write:

$$n_1 sin\theta_1 = (n_1^2 - n_0^2 cos^2\theta_0)^{1/2} \qquad (11)$$

2.3 The ideal interface and the Fresnel reflectivity curve

Let us consider here a neutron beam propagating in vacuum (or air) and reflecting on an ideal planar substrate without any layer at the surface. If the interface is perfect with no roughness, such a system is a diopter. The reflectivity of such an interface is called the Fresnel reflectivity (R_F). The Fresnel reflectivity depends on the SLD of the substrate ρ_s and writes:

$$R_F = \left|\frac{B}{A}\right|^2 = \left|\frac{q-q_s}{q+q_s}\right|^2 = \left|\frac{1-\left(1-\left(\frac{q_c}{q}\right)^2\right)^{1/2}}{1+\left(1-\left(\frac{q_c}{q}\right)^2\right)^{1/2}}\right|^2 \qquad (12)$$

where q_c and q_s are obtained from:

$$q_c^2 = 4\pi\rho_s \qquad (13)$$

$$q_s^2 = q^2 - 4\pi\rho_s = q^2 - q_c^2 \qquad (13')$$

q_c is the critical scattering vector that separates total reflection that occurs below q_c, from partial reflection where a part of the beam is refracted (above q_c). At q_c, the wave propagates only at the surface and parallel to it. Far from the critical angle ($q >> q_c$), $R_F \propto (1/q)^4$. This means that the decay in reflectivity is always very fast whatever the layer at the surface. The NR curves R = f(q) are thus traditionally represented in logarithmic-linear scales or logarithmic-logarithmic scales (Figure 2). Another convenient way to represent R_F is the *Fresnel representation f*(q) = R(q)q^4 for which the q^4 term compensates the intrinsic q^{-4} decay of the NR curve to highlight the features coming from the layers at the interface.

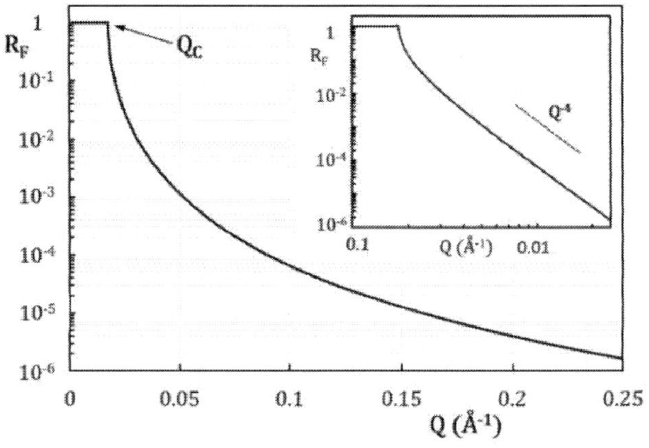

Fig. 2. Logarithmic-linear representation and logarithmic-logarithmic representation (inset) of the Fresnel reflectivity as a function of Q for ρ_s = 6.38 10^{-6}Å^{-2} (air/D_2O interface).The SLD of air ρ_{air} is considered as equal to 0 owing to the very low density of the gas.

2.4 The case of a homogeneous layer on a substrate

Let us consider a single homogeneous layer of thickness d deposited on the surface of a perfectly flat substrate of thickness that can be considered infinite (Figure 3). Such a system is composed of three media: the air (of refractive index 1 and scattering length density 0), the layer (of refractive index n_1 and scattering length density ρ_1) and the substrate (of refractive index n_s and scattering length density ρ_s) and of two interfaces. Let us set the air/layer interface at $z = 0$ and the layer/substrate interface at $z=d$. An incident neutron beam coming from the air will be partially reflected at the first interface ($z = 0$) and partially transmitted (for $q > q_c$). The transmitted wave will then be partially reflected and partially transmitted at the layer substrate interface ($z = d$). The two reflected waves will interfere constructively or destructively depending on their phase difference which is related to their difference on optical path and thus to the thickness of the layer. It therefore appears qualitatively that the reflectivity curve will exhibit interference fringes that enable the measurement of the layer thickness.

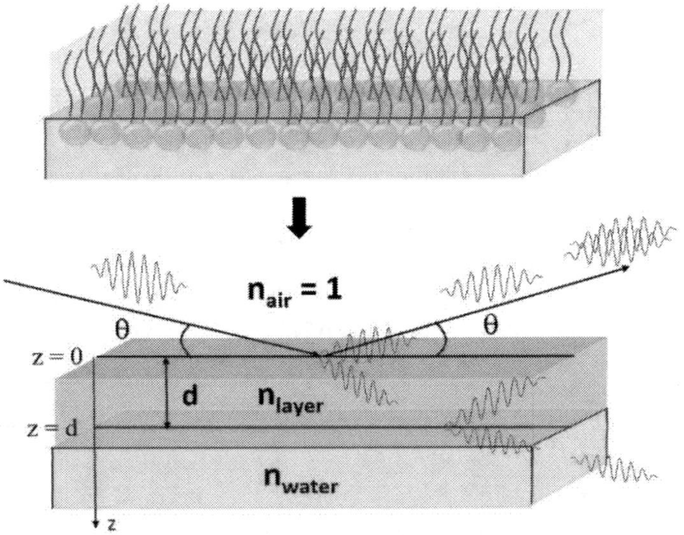

Fig. 3. (Color online). Reflection of a homogeneous monolayer of phospholipids at the air/water interface. The hydrophobic tails form a thin film of homogeneous SLD and thickness *d*. The water is semi-infinite substrate. The phospholipids heads are not considered here.

Figure 4 presents some typical NR curves for monolayers of different thicknesses and SLD at the air/water interface. R exhibits typical oscillations that are called the Kiessig fringes. The fringe spacing is $\boldsymbol{\Delta q = 2\pi/d}$. This is the Bragg relationship corrected from the critical angle:

$$2d\sqrt{\sin\theta^2 - \sin\theta_c{}^2} = m\lambda \tag{14}$$

where m is an integer.

The inter-fringes distance is linked to the thickness of the layer and their amplitude comes from the difference of SLD between the monolayer and both air and substrate.

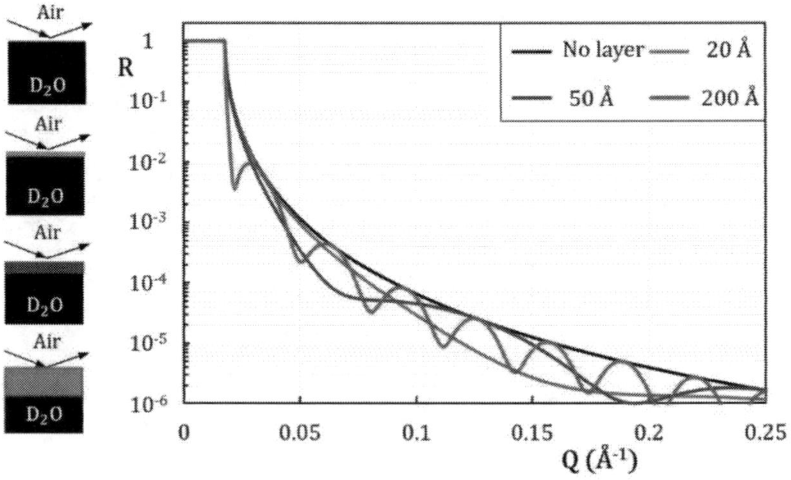

Fig. 4a. (Color online). Reflectivity R calculated for a single layer at the air/D_2O interface of different thicknesses *d* and constant ρ_{layer} (4.6 10^{-6} Å^{-2}). The sketches in the left panel are schematic representations of the system.

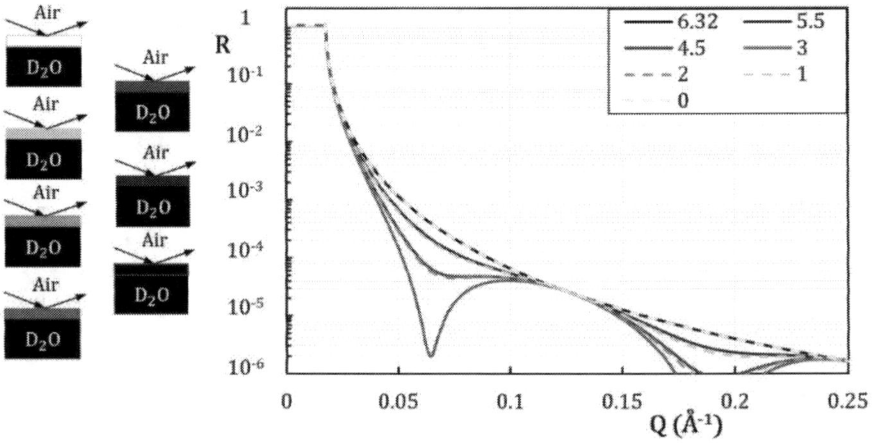

Fig. 4b. (Color online). Reflectivity R calculated for a single layer at the air/D_2O interface for different SLD ρ_{layer} (x 10^{-6} Å^{-2}) and constant thickness *d* (50 Å). The sketches in the left panel are schematic representations of the system.

2.5 Roughness and inter-diffusion

In the previous parts, we have considered that the interface between the layer *p* and the layer *p+1* was perfect with no heterogeneities, which means that there is a sharp discontinuity of SLD when passing from ρ_p to ρ_{p+1} (step function). This is only theoretical; in practice, interfaces are never so abrupt because of two different physical phenomena: (*i*) roughness and/or (*ii*) inter-diffusion (see Figure 5). Roughness is due to the fact that the surface is irregular, as shown in Figure 5.a. and is equivalent to the case where neutrons are reflected by many surfaces with different angular orientations and a significant part of them is out of the specular plane. Inter-diffusion occurs when two successive layers of different composition diffuse slightly one into each other, creating a smooth transition (Figure 5.b).

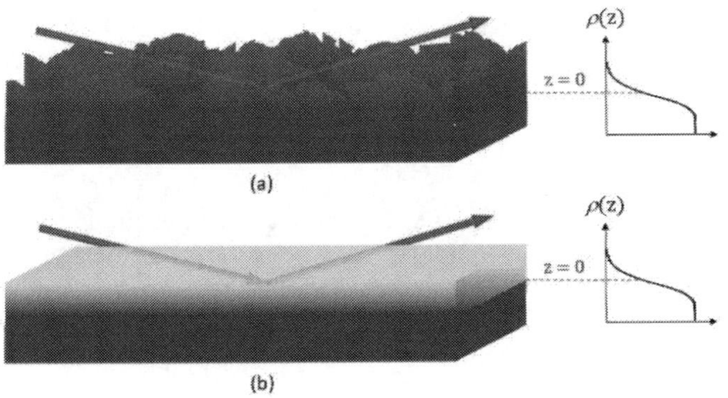

Fig. 5. (a) Schematic representation of roughness; (b) schematic representation of inter-diffusion. The profiles represented on the right show the effective ρ(z) profile averaged on the whole surface.

Assuming that both kinds of interfaces have a Gaussian profile, they will be simulated by the same error function, where *p*/*p*+*1* is the interface between two successive layers:

$$erf\left(\frac{z-z_{p/p+1}}{\sigma_{p/p+1}}\right) = \frac{2}{\sqrt{\pi}}\int_0^{(z-z_{p/p+1})/\sigma_{p/p+1}} e^{-t^2}\,dt \qquad (15)$$

This function has an inflection point at $z_{p/p+1}$. $\sigma_{p/p+1}$ is the inverse of the tangent slope of the curve at $z_{p/p+1}$. The thickness of the interface is given by $2\sigma_{p/p+1}$. Another method to take into account roughness is to discretize the density profile in a large number of layers having a low variation of ρ from one layer to the next. However, the calculation time may become very long.

Figure 6 shows the effect of roughness/interdiffusion on the NR curve calculated in the case of a single layer at the air/water interface (same parameters as in Figure 4).

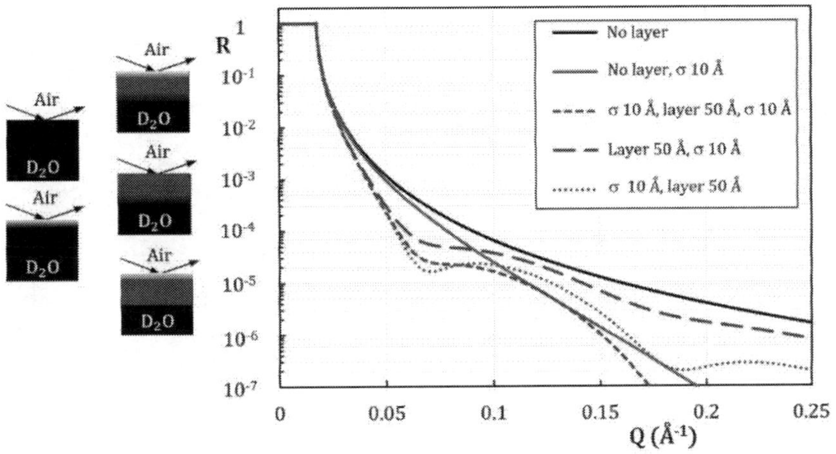

Fig. 6. (Color online). Influence of the roughness (σ = 10 Å) on the reflectivity R calculated for a single layer (ρ_s = 4.6 10^{-6} Å^{-2}, *d* = 50 Å). The sketches in the left panel are schematic representations of the system.

3 Practical aspects

This part is devoted to the experimental aspects of NR to help users achieving successful experiments.

3.1 Instruments

3.1.1 Principle of measurement

In neutron sources, whether neutrons are produced with a continuous flux in a nuclear reactor or by pulses in a spallation source, the energy of the produced neutrons is of the order of a few MeV; they must, therefore, be thermalized in order to perform neutron scattering experiments. The neutron energies needed for NR are of a few meV (cold neutrons) with corresponding wavelengths ranging between 2Å and 20Å. The thermalized neutron beam is always polychromatic.
It exists two ways to measure the reflectivity of a sample as a function of the scattering vector $Q = 4\pi\sin\theta/\lambda$: either by varying θ at fixed λ, or by varying λ at fixed θ.

In the first case, called **θ -2θ mode**, a monochromator, *e.g.* a mono-crystal, is used to select the needed wavelength λ; the neutron beam is then collimated using collimation slits and sent to the sample at grazing incidence. One then measures the reflected beam, step by step, by rotating the sample of an angle θ and the detector of an angle *2*θ. The illumination of the sample varies from one angle to the other, which has to be taken into account in data reduction.

In the second case, called **time-of-flight mode**, one exploits the fact that neutrons travel at a finite speed v that is rather slow as it lies between a few 100 m/s and a few 1000 m/s for cold neutrons. Neutron speed v is related to wavelength λ through the de Broglie relationship $v=h/m\lambda$, where h is the Planck constant and m the neutron mass. In a time of flight experiment, a pulsed white beam is sent to the sample at a fixed angle θ: it is possible to separate neutrons of different wavelengths through their speed, the fastest neutrons arriving on the detector before the slower ones. Pulses are spontaneously generated in a spallation source but have to be created by a chopper, a rotating disk containing neutron-transparent windows, in a reactor source. The collimated neutron beam is sent at grazing incidence on the sample; if one defines the origin of times $t_0 = 0$ the time at which the chopper generates a pulse, the wavelength of neutrons arriving at the detector, placed at a known distance L, is determined by the measurement of the time t taken to every neutron to *flight* from the chopper to the detector, since $\lambda=th/mL$. It is assumed that the speed of neutrons does not change when they interact with the samples (elastic scattering). Measuring the coefficient of reflection is simple: the white incident polychromatic beam is first measured directly on the detector without any reflection and the measured reflected beam is then divided by such incident white beam for every λ value.

For biological and soft matter studies, the dedicated reflectometers work with the time-of-flight mode because it has incomparable advantages with respect to the θ -*2*θ mode: (*i*) working at constant θ does not force to rotate the sample, which is highly appreciable for measurements at the air/water interface; (*ii*) all the points of a reflectivity curve are measured at the same time, which removes any incertitude on the possible ageing of the sample during measurement; (*iii*) kinetic measurements are possible.

In a NR experiment, the incident angle on the surface under study is always very small, ranging typically from 0.5° to 5°. Measurements at lower angles are indeed practically almost impossible because the reflected beam has to be separated from the incident non-deviated direct beam. The range of scattering vectors lies then between 0.003 $Å^{-1}$ and 0.3 $Å^{-1}$ for cold neutrons. This fixes the typical range of thickness that can be measured from around 5 Å to 2000 Å. The collimation is achieved by a simple set of rectangular slits made of neutron absorbers of various apertures. The sample is mounted on a goniometer in order to set the desired angle. In the experiments at the air/water interface where it is not possible to rotate the sample, the neutron beam is deviated using a neutron mirror upstream of the sample in order to obtain the desired angle at the air/water measurements. A typical sketch of a time-of-flight reflectometer at the air/water interface is depicted in Figure 7.

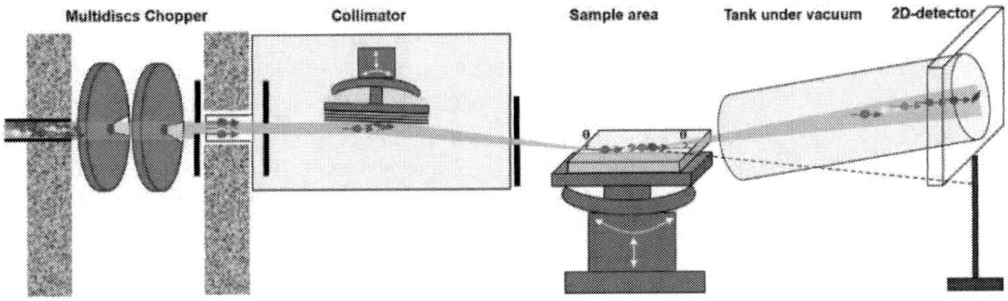

Fig. 7. (Color online). Typical sketch of a time-of-flight reflectometer on a continuous neutron source (reactor). A double-chopper located in the casemate creates pulses of polychromatic neutrons from the continuous beam. The chopped beam is then collimated by slits and can be deviated by a supermirror if necessary for measurements at the air/water interface. The reflection of neutrons on the sample is considered as elastic. The determination of the speed, *i.e.* of the wavelength, is done by measuring the time of neutron flight from the chopper to the detector. Neutrons of different wavelengths are depicted by arrows of different colors.

The typical acquisition time depends on the requested minimal reflectivity *R*. If *R* is of the order of 10^{-5}, a few minutes are enough to obtain a correct statistics on state-of-the art instruments with a sample of ~10 cm² (see next section), but this time has to be extended up to a few hours to reach 10^{-6} or below.

3.1.2 Experimental resolution

The major weakness of NR is the poor flux of the neutron beam provided by the actual sources, whether reactor or spallation sources, even on the most powerful ones. In order to make a proper measurement with a very good signal-to-noise ratio, a compromise has to be done between flux and resolution. In biological or soft matter studies, performing the experiments in degraded resolution is not an issue since the probed interfaces are soft. The usual chosen $\delta q/q$ resolution lies thus generally between 5% and 10%. This resolution originates both from the $\delta\theta/\theta$ of the collimation and from the $\delta\lambda/\lambda$ of the chopper (or a monocrystal or a mechanical selector in the case of a reflectometer working in θ-2θ mode with a monochromatic beam).

$$\left(\frac{\delta q}{q}\right)^2 = \left(\frac{\delta\theta}{\theta}\right)^2 + \left(\frac{\delta\lambda}{\lambda}\right)^2 \qquad (16)$$

The divergence of the incident beam $\delta\theta$ and the angular resolution $\delta\theta/\theta$ are set by the apertures of the entrance and exit slits of the collimator, which are opened usually at around 1 or few mm, and by the collimator length, which is usually a few meter long. For an entrance slit of aperture S_{ent}, an exit slit of aperture S_{exit} and a collimator length D_c, the divergence is $\tan\delta\theta = (S_{ent}+S_{exit})/2D_c$; it can be taken into account in the calculation. The expressions for

https://doi.org/10.1051/epjconf/202023604001

reflectivity presented in the precedent part have been calculated for an incident beam without divergence. The calculation of the $\delta\lambda/\lambda$ of a chopper is much trickier and is detailed in [13].

3.2 Samples and possible interfaces that can be probed

The geometry of a NR experiment imposes three imperative conditions that have to be absolutely fulfilled to make it feasible. First, the medium in which the neutron beam is travelling upstream the interface and after reflection has to be totally or partially transparent to neutrons, otherwise the signal is lost. Second, the flatness of the studied interface has to be as best as possible: if the interface has a curvature, this would have an effect equivalent to an increase of the divergence $\delta\theta$ of the beam. Although this divergence could be taken into account by an increase of the effective $\delta\theta$ within the resolution function during the data modelling, it has to be as limited as possible. This imposes some limitations on the different possible geometries of measurements and on the samples themselves. Third, the roughness of the surface has to be very limited, otherwise it would induce a very steep drop of reflectivity that leads to a loss of information on the SLD profile.

3.2.1 The different possible geometries of measurements

In biology, almost all NR studies focus on processes that occur between biological species (proteins, viruses, peptides, …) and phospholipids bilayers (membranes) or monolayers. The most interesting interfaces are then **liquid/liquid interfaces and air/water interfaces**. However, there are only very sparse studies probing true liquid/water interfaces by NR since this is something hardly challenging for some reasons that are detailed elsewhere [14]. Thus, in practice, liquid/water interfaces are replaced by experiments at **solid/water interfaces** where membranes are deposited on solid substrates, as depicted in the book section written by Gerelli.

Experiments at the air/water interface are very easy to implement since air is almost transparent to neutrons and therefore the attenuation of the neutron beam on a few *cm* in air is negligible.

Conversely, in NR experiment at the solid/water interface, the transparency of media in which the neutron beam has to travel is a key point to consider. In practice, two processes attenuate the beam: absorption and incoherent scattering. In biological studies, neutron absorption is never an issue because water and biological compounds contain none of the few elements of the periodic table that have an important absorption cross-section for neutrons (^{3}He, ^{6}Li, B, Gd, …) [11]. However, the beam transmission can also be reduced due to incoherent scattering, that occurs with a constant probability over 4π independently from the position of scatterers [9], when this latter is important. Indeed, although the incoherent scattering cross-section is null or very weak for most elements of the periodic table, it is unfortunately very high for hydrogen ^{1}H and important for ^{2}H deuterium. For example, the transmission of a neutron beam is only 0.5 through 1 mm of H_2O and 0.8 through 2 mm of D_2O for 6 Å neutrons. Therefore, since in a NR experiment the beam has an incident grazing angle (less than 5°), the neutron beam has to travel for several centimeters within the sample; in this case a particular geometry has to be used in order to avoid crossing water, as shown in Fig.8.c. The neutron beam has thus to pass through the solid support that has to be made with a material transparent to neutrons.

3.2.2 Samples and samples environments

The surface of the sample has to be optimized. Since the raw number of reflected neutrons is directly proportional to the surface of illumination of the sample by the neutron beam prior to the data reduction step, the surface of the sample has to be maximized in order to get a good signal-to-noise ratio with reasonable acquisition times. With the current performances of the state-of-the-art reflectometers, the order of magnitude of the sample surface allowing good measurements is a few cm^2. It must be avoided to decrease surfaces area lower than ~ 1 cm^2. Moreover, at the air/water interface, it is mandatory that the sample surface is sufficiently large so that the beam illuminates a perfectly flat area of the surface, thus located away enough from the meniscus at the edges of the trough containing the water.

At the solid/liquid interface, the solid substrate to be crossed by neutrons must be transparent to neutrons, must have a very small (root mean square (*rms*) roughness and must be rigid enough not to deform mechanically under internal stresses, so that it does not create a curvature at the interface. In practice, the only solid substrates that are used in NR for biological studies are silicon wafers (Fig. 8.d). Sapphire blocks or quartz crystals have also been seldom used in soft matter but they are much more expensive. Silicon wafers are indeed nearly transparent to neutrons because their incoherent scattering cross-section is null and their absorption cross-section is very weak. They can easily be purchased with an excellent mosaicity (*rms* roughness of a few Å) at low cost from silicon manufacturers. They are always covered by a very thin layer of SiO_2 (~ 10 to 20 Å) at their surface. Such silica layer may pose problems during data fitting but they can also offer an advantage since Si-OH groups enable many potential chemical modifications in order to render the surface hydrophilic, hydrophobic, or electrically charged, for example.

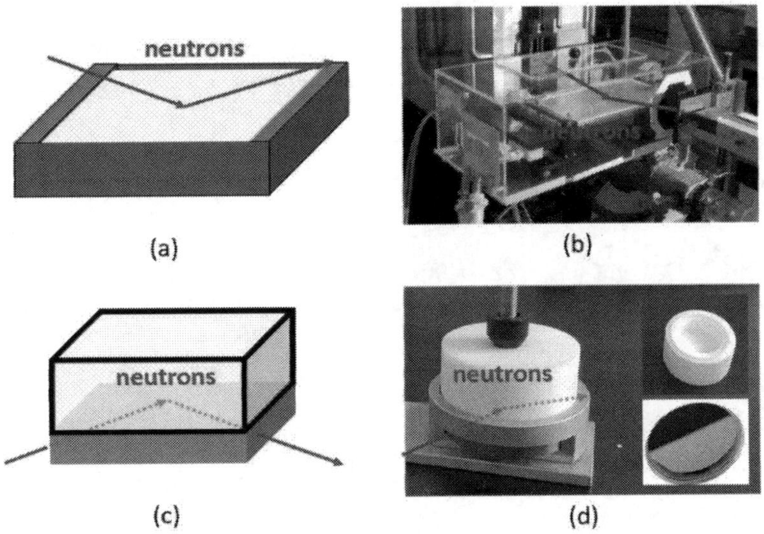

Fig. 8. (Color online). (a) Standard measurement geometry at the air/water interface. (b) Langmuir trough installed on the EROS reflectometer (LLB, France) [15] for measurements at air/water interface. (c) Standard measurement geometry at the solid/water interface. (d) Typical cell for measurement at solid/water interface; insets show respectively the reservoir for water (upper picture) and the silicon wafer with its polished surface (lower picture).

Thanks to the fact that neutrons are poorly absorbed by almost all atoms of the periodic table, it is easy to design various sample environments for *in situ* experiments. At the air/water interface, the most used equipment is the Langmuir trough (Fig. 8.b) since it enables to set the concentration of the monolayer under study at the targeted surface pressure. At the solid/water interface, there are various measurements cells in the different neutron facilities, that can be equipped with various ancillary options like solvent circulation baths, temperature controllers or electrochemical cells for electric measurements [16]; an example of the application of an electrochemical cell to lipids is presented in [17]. When filling such cells, it has to be avoided that air bubbles locate at the silicon/water interface, which is easily achieved on horizontal reflectometers when the water is placed above the silicon (Fig. 8.d).

3.3 Access to instruments

Reflectometers are accessible to the scientific community in large neutron facilities. The beamtime (*e.g.* 2 to 5 days) is allocated after acceptation of proposal experiments after reviewing by a committee of experts. Call for proposals takes place generally twice a year. Fast access or test experiments are also sometimes possible and have to be negotiated with facilities. The actual landscape of European neutron facilities has strongly changed these last years with the closure of Orphée reactor at Saclay (France) and HMI at Berlin (Germany). State-of-the-art reflectometers can now only be accessed in Europe at ISIS (United Kindgom), MLZ (Germany), PSI (Switzerland) and at ILL, the European source located in Grenoble, France [18]. Two reflectometers are especially suited to biological studies at ILL, namely FIGARO, a horizontal reflectometer dedicated to soft matter and biophysics, and D17, a vertical reflectometer for multipurpose measurements. In the near future, the new European Spallation Source (ESS, Lund, Sweden) will be operating and FREIA, a high flux horizontal reflectometer, is forecast for 2025 in the instrumental suite [19].

The authors thank the organizers of the "Neutrons and Biology School 2018" of the *French Neutron Society* for their kind invitation to participate to this special issue. They also thank D[r] Louis Bondaz whose *PhD* manuscript has inspired Figures 2, 4, and 6 of this course.

References

1. M. Wolff, EPJ Web Conf., **188**, 04002 (2018)
2. J. Penfold, R. K. Thomas, J. Phys. Condens. Matter, **2(6)**, 1369–1412 (1990).
3. T. P. Russell, Mater. Sci. Reports, **5**, 171–271, (1990)
4. L. Braun, M. Uhlig, R. von Klitzing, R. A. Campbell, "Adv. Colloid Interface Sci., **247**, 130–148 (2017)
5. H. P. Wacklin, Curr. Opin. Colloid Interface Sci., **15**, 445–454, (2010)
6. G. Fragneto, Eur. Phys. J. Spec. Top., **213**, 327–342 (2012)
7. G. Fragneto-Cusani, J. Phys. Condens. Matter, **13(21)**, 4973–4989 (2001)
8. F. Ott, École thématique la Société Française la Neutronique, **13**, 02004 (2014)
9. G. L. Squires, *Introduction to the Theory of Thermal Neutron Scattering*, (Cambridge: Cambridge University Press, 1978)
10. V. Sears, Neutron news, **3**, 26–37 (1992)
11. https://www.nist.gov/ncnr/neutron-scattering-lengths-list.
12. L. G. Parratt, Phys. Rev., **95(2)**, 359–369 (1954)
13. D. Lairez, A. Chennevière, F. Ott, J. Appl. Cryst, **53,** 464-476 (2020)
14. F. Cousin, A. Chennevière, EPJ Web Conf., **188**, 04001 (2018)
15. F. Cousin, F. Ott, F. Gibert, A. Menelle, Eur. Phys. J. Plus, **126**, 1–11 (2011)
16. A. Koutsioubas, D. Lairez, G. Zalczer, F. Cousin, Soft Matter, **8**, 2638–2643 (2012)
17. G.C. Fadda, D. Lairez, Z. Guennouni, A. Koutsioubas, Physical Review Letters, **111(2)**,

028102 (2013)
18. https://www.ill.eu/
19. https://europeanspallationsource.se/

Applications of neutron reflectometry in biology

Yuri Gerelli[1,*]

[1]Polytechnic University of Marche, Department of Life and Environmental Sciences, Via Brecce Bianche, Ancona, Italy

Abstract. Over the last 10 years, neutron reflectometry (NR) has emerged as a powerful technique for the investigation of biologically relevant thin films. The great advantage of NR with respect to many other surface-sensitive techniques is its sub-nanometer resolution that enables structural characterizations at the molecular level. In the case of bio-relevant samples, NR is non-destructive and can be used to probe thin films at buried interfaces or enclosed in bulky sample environment equipment. Moreover, recent advances in biomolecular deuteration enabled new labeling strategies to highlight certain structural features and to resolve with better accuracy the location of chemically similar molecules within a thin film.
In this chapter I will describe some applications of NR to bio-relevant samples and discuss some of the data analysis approaches available for biological thin films. In particular, examples on the structural characterization of biomembranes, protein films and protein-lipid interactions will be described.

1 Introduction

The use of neutron reflectometry (NR) in biological sciences has been growing enormously in the last decades. In the field of soft matter and biophysics, the technique was introduced in the '80s for the study of surfactant thin films but NR, as a technique, remained under exploited for biological relevant samples. In the last two decades, use of NR in the filed of biological sciences increased regularly [1]. This was possible mainly because of improvements in instrumentation, sample environment and modeling approaches. These developments allowed investigations of systems of increasing complexity. The fact that NR is a technique applied to interfaces limits the type of samples that can be studied. However, interfaces are ubiquitous in nature and they play important roles in all the living organisms. The main example of a natural interface is represented by the cell membrane, the active barrier separating the interior of cells from the extracellular environment. Because of their nanometric thickness and macroscopic lateral size, cell membranes are a perfect sample for NR experiments. In fact, they represent one of the largest area of application of NR in biology. The first systems investigated were simple solid-supported lipid bilayers (SLBs) composed by a single molecular species and they served as a benchmark to develop sample preparation techniques and analysis methods. Nowadays, SLBs of increased complexity, formed by synthetic or natural molecules, including proteins can be prepared and investigated by NR as it is described in this chapter.

*Corresponding author: y.gerelli@univpm.it

Another bio-relevant system to which NR was successfully applied is represented by protein films. Their formation at interfaces is of multiple interest. For example, at the macroscale, engineering of protein coatings is essential to improve the osseointegration of medical implants. At the nanoscale, the formation of protein films around colloidal particles is fundamental in drug-delivery, therapeutic and diagnostic applications. In fact, when nanoparticles are administered into biological fluids they get covered by the so-called protein corona. This protein film represents the interface to the human body and therefore it regulates nanoparticles systemic circulation, biodistribution, and bioavailability.
Finally the combination of proteins and cell membranes represents probably the most investigated class of bio-relevant systems for NR experiments. In fact, interactions of proteins or peptides within and at the surface of cell membranes are fundamental for the understanding of molecular processes involved in diseases, for the development of new drugs and for the design of novel anti-microbial strategies [2–4].
For the study of the above mentioned systems, a great advantage of NR with respect to many other surface-sensitive techniques is a labeling strategy based on the isotopic substitution [5]. This method is referred as *contrast variation* and in its most common application it exploits the substitution of the naturally occurring hydrogen atoms (protium) by deuterium ones. Lipids [6–8], proteins [9] and many other biomolecules can be either synthesized or naturally produced by microorganisms in partially or fully deuterated forms. The use of hydrogenated and deuterated molecules in the same samples allow to selectively locate individual molecular species and it is therefore particularly suitable for the structural determination of complex biological systems.
Within the framework of biological application of NR, I will briefly describe the main sample preparation approaches currently available and the analysis methods that can be exploited for the analysis of extremely flat and homogenous or rough and diffuse biological thin films. In the second part of the chapter I will give a series of examples to highlight what type of systems can be studied and what information can be obtained.

2 Sample preparation approaches

As described by Cousin and Fadda in the chapter "An introduction to neutron reflectometry", NR requires the presence of an interface onto which samples, in the form of thin films, can be deposited or adsorbed. Typical interfaces for biological samples are the solid-liquid, liquid-air and solid-air. At these interfaces samples can be deposited by several techniques such as adsorption from a liquid phase, drop-casting and spin-coating, Langmuir-Blodgett and Langmuir-Schaefer deposition methods [10, 11].
Adsorption from a liquid subphase is a well established method to create self-assembled films at an interface. Among several examples it might be worth mentioning the case of vesicles fusion, a technique used to obtain a single lipid bilayer at a solid-liquid interface [12] or the adsorption from bulk in which a molecular film is created by the natural tendency of the molecules of interest in partitioning between the aqueous phase and the interface [13, 14]. In both cases the bulk liquid phase is enriched with the molecules of interest and, as a result of a kinetic process, these molecules can self assemble into an interfacial thin film.
Spin-coating is another well established technique to deposit in a controllable way molecules at a solid-air interface. A wanted amount of a solution containing the molecules of interest can be added drop-wise onto a solid substrate which rotates at a controlled rate. While the solvent is either ejected as an effect of the rotation or evaporates, the molecules it contains can self-assemble or organise at the interface. This approach is typically used to create multilayered structures [15, 16] that can be then either studied at the solid-air interface or exposed to an aqueous phase in solid-liquid geometry.

Langmuir-Blodgett and Langmuir Schaefer are two deposition techniques in which monolayers prepared at the air-water interface (typically using an equipment called Langmuir trough) can be transferred onto solid substrates. These techniques are widely used to prepare single solid supported bilayers for which is crucial to control the composition of the individual monolayers, as in the case of asymmetric systems [17, 18]. In addition, these two techniques are essential to prepare lipid and surfactant mono- or multi-layer systems as well as double or multi-bilayers systems with an accurate control of the number of bilayers and of their composition [19, 20].

3 Modeling method for biological problems

One of the main bottleneck limiting the use of NR in biological sciences is the necessity of adapting or developing new modeling approaches that could be used to describe the structure of samples of increasing complexity. In the following I will review a series of modeling approaches that can be used for the analysis of biological systems at interfaces. All of them are based on the computation of a theoretical scattering length density (SLD) profile describing the distribution of molecular groups or atomic species along the vertical direction with respect to the interface. The distance from this interface is typically indicated as z and expressed in Ångströms or nanometers. In general, from an SLD profile it is possible to compute the corresponding theoretical reflectivity curve, $R(Q)$, to be refined against the experimental data. This leads, in turns, to the refinement of the SLD profile that can be therefore interpreted to obtain a correct description of the structure of the sample. In fact, since the SLD is determined by the number density of atomic nuclei (see chapter "An introduction to neutron reflectometry" for the definition of SLD), these profiles are directly linked to the distributions of the components in the sample and therefore describe the vertical structure of interfacial thin films. By applying the contrast variation method, the scattering signal of one or more components can be altered resulting in altered SLD profiles. In this way, the location of selected components within the sample can be highlighted.
In the following I will briefly describe the approaches that are commonly used for the modeling of NR data of bio-relevant systems.

3.1 Slab models

The slab model is the oldest model developed for the analysis of NR data which can be explicitly used in the calculation of reflectivity from stratified interfaces [21, 22]. In this model a thin film is described as discrete series of layers. In this context, a layer is defined as a portion of the sample characterized by a SLD value that differs from those of the neighbouring regions along the z direction. An interface is therefore the line in space where a change in SLD takes place. Since reflectometry probes only the vertical direction with respect to an interface, SLD changes must occur along the z sample axis while lateral homogeneity is preserved. This implies that even if the sample is physically not stratified (for example it is composed by a single molecular layer) it might be modeled as a series of slabs if the SLD(z) changes within the sample. This is, for example and among many others, the case of a surfactant layer at the air-water interface [23] or of a supported lipid bilayer (see Figure 1).
In general, each slab is characterized by three structural parameters, namely thickness, SLD and interfacial roughness. In some cases, especially upon the use of the contrast variation technique [5], the total SLD of a given layer can be redefined as a function of the SLD of the molecules composing the layer and of the volume fraction of solvents present in the same region. This redefinition is commonly used in the case of thin films in which hydration plays

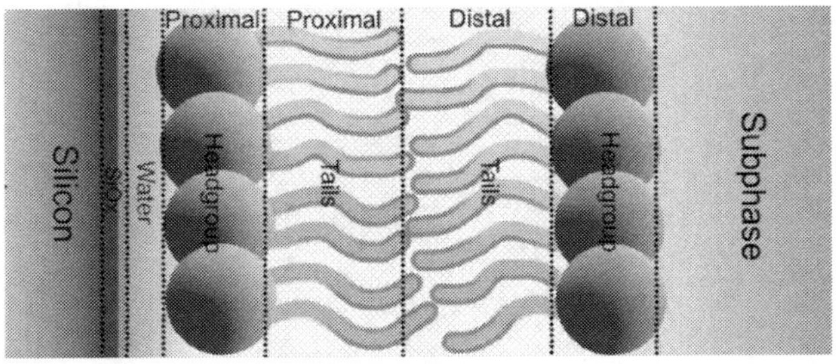

Figure 1. Schematic representation of the slab model used to describe a SLB on the top of a silicon substrate. Reprinted with permission from Ref. [24].

an important role [25] and for mixed systems [8]. In general, the total SLD ρ of a slab composed by n different chemical groups or materials can be expressed as

$$\rho = \sum_{i=1}^{n} \Phi_i \times \rho_i \tag{1}$$

where Φ_i ($\sum_{i=1}^{n} \Phi_i \equiv 1$) is the volume fraction of the i-th component and ρ_i is its SLD. Equation 1 is particularly useful in case of hydrated samples where one of the components represents hydration water.

The slab model is rather general and can be applied to a large variety of systems. It is, in fact, implemented in the most common software applications devoted to the analysis of NR data originated from soft or biological samples [25–29]. However, the slab model might fail if the interfacial roughness between subsequent slabs is too large (approximately larger than the slab thickness) or if the sample is laterally and vertically inhomogeneous. In these cases, alternative approaches should be used and some of them are described in the following sections.

3.2 Functional component groups

As reported in the previous section, the main limitation of the slab model is the description of (vertically, along z) diffuse layers. This can be the case, among several others, of polymer brushes deposited onto surfaces [30–32] or of inhomogeneous systems [33], for which a description through a limited number of well defined layers might fail. As reported in the cited literature, these systems can be modeled by a combination of slabs, for the homogeneous regions (if any) of the sample, and by functional distributions, describing volume fraction profiles, for the inhomogeneous ones. In many cases, the shape of the functional profiles to be used can be retrieved from theoretical predictions while their parameters can be varied during the data analysis. For example, a lipopolysaccharide (LPS) monolayer deposited at the air water interface can be modeled as described in Figure 2, where the polysaccharide chains are described by a stretched exponential function [33] while the hydrocarbon tails and the oligosaccharide core are described sufficiently well by two slabs. The same approach was used for a large variety of systems [30–32] in which at least one component can be described as a polymer brush. Volume fraction profiles can be converted into SLD profiles simply by weighting them according to the nominal SLD values associated to the desired material or chemical species. Similarly, from a given volume fraction profile several SLD profiles can be obtained to simulate the effect of contrast variation i.e. the change of a given SLD value

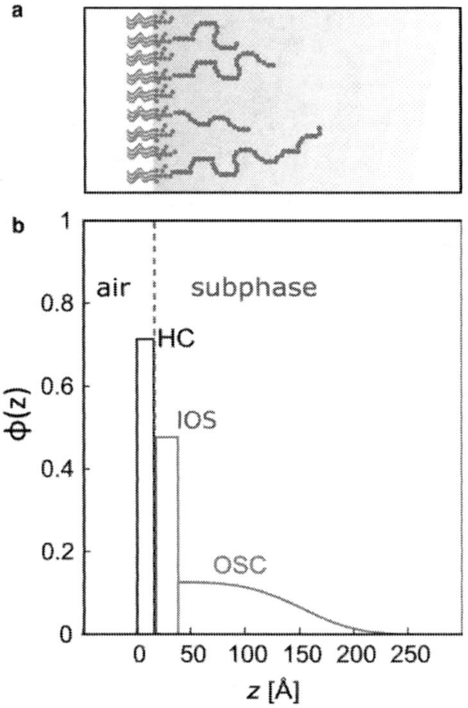

Figure 2. (a) Schematic representation of a lipopolysaccharide monolayer at the air-water interface. (b) Theoretical model used to describe the volume fraction distributions of the LPS groups at the interface. The components are hydrocarbon chains (HCs), inner oligosaccharides (IOSs) and O-side chains (OSCs). Reprinted from Biophysical Journal, 116, S. Micciulla, Y. Gerelli and E. Schneck, Structure and Conformation of Wild-Type Bacterial Lipopolysaccharide Layers at Air-Water Interfaces, 1259–1269, Copyright (2019), with permission from Elsevier.

taking place because of isotopic substitution. Once the SLD profiles are generated, the optimization procedure of the structural parameters is exactly the same as the one described for the slab model. In fact, to calculate the corresponding reflectivity curve the functional SLD profile is sliced in finite-size slabs (typically small enough to reproduce accurately the original curve) and used within the standard Parratt's or Abelès' formalisms.

A similar approach was also developed for the analysis of SLBs, not because of the failure of the slab model but because of the necessity to compare models obtained from NR measurements on flat membranes to those obtained from vesicle samples by small angle X-ray and neutron scattering data [34–36]. In this approach, a lipid molecule is modeled as the sum of the probability distributions of its component groups calculated along the z direction (parallel to the lipid acyl chains). The components could be (for example): choline methyl groups, phosphate + CH_2CH_2N, carbonyl-glycerol (GC), hydrocarbon methylene groups (CH_2) and methyl groups of hydrocarbon chains (CH_3). Such a model for a SLB at a solid-liquid interface is given in Figure 3. In this case, all the groups are represented by functions whose parameters can be adjusted, within the appropriate constraints, and refined during data analysis. The only component not described by an analytical function is the probability distribution of water, that is calculated as the complement to unity of all the other probabilities. This allows the model to reproduce not only perfect SLBs but also systems with a non uniform and partial coverage. The SLD profile is obtained by the sum of all the distribution functions

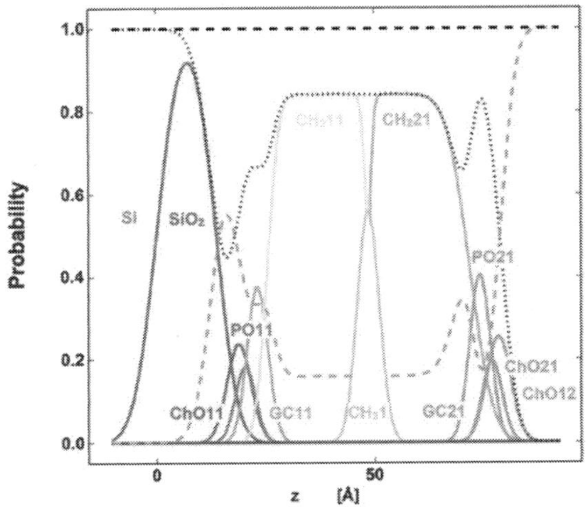

Figure 3. Scheme of component group probability distributions within the SLD profile model of a supported lipid bilayer. The components are: bulk silicon – Si, a silicon dioxide layer – SiO_2, hydrocarbon chain region CH_2, water – H_2O choline – ChO, phosphate – PO, carbonylglycerol – GC and methyl groups – CH3). XY in the group names label are references for the leaflet in which the groups are. Reproduced from Ref. [36] with permission from the Royal Society of Chemistry.

weighted by the nominal SLD values associated to a given molecular group. This profile can be sliced into finite-size slabs and reflectivity calculated as described in section 3.1. Examples of applications of these approaches to biological systems are given in Section 4.2.2.

3.3 Simulation-based methods

In soft matter, biophysics and biology, the use of simulations for the analysis of large scale structures measured by means of neutron scattering experiments is well established for most of the scattering techniques available [37–40]. In the context of neutron reflectometry applied in biological sciences only in the last two decades, the use of MD simulations to support the interpretation of experimental data gained a reasonable traction. As already mentioned in the previous sections, the result of an analysis of NR data consists in scattering length density or in volume fractions profiles. Similar profiles can be obtained from the output of computer simulations and therefore a comparison between experimental and computational results is rather straightforward. The use of computer simulations to complement or cross-validate the analysis of experimental NR data has been reported for several systems, including proteins at interfaces [41], membranes and peptide-membrane interactions [28, 42, 43].
More recently, different groups have started to use MD simulations not only as a complementing information but also as a tool for the analysis of NR data. This was possible because of the improvements made in the quality of parameters sets in MD simulations. An example of the importance of the force-field for a correct reproducibility of experimental data has been given in the work by Koutsioubas on the phase transition of a solid-supported lipid bilayer [44]. One of the first remarkable examples of combination of traditional modeling approaches and numerical simulation was described by Heinrich and Lösche [45]. In their work, they demonstrated that it is possible to determine the out-of-plane localization of transmembrane proteins in lipid membranes with sub-nanometer precision by using all-atoms simulations to interpret NR data.

In a more recent work, McCluskey *et al.* [46] investigated whether the use of atomistic simulations are necessary to reproduce accurately experimental reflectometry data or whether coarse-grained representations are sufficient. By the comparison between the results obtained using different MD potential models and those originated from a traditional slab description of a monolayer sample, the authors concluded that discrepancy between simulations and experimental data are likely to be observed if the coarse-grained representation is too simplified. As the resolution of the potential model increased, the agreement with experimental data improved. However, any coarse-grained representation could describe with the same goodness as a traditional slab model, the reflectivity curves.
Concerning bio-systems, Darré *et al.* made a direct comparison not only between SLD profiles calculated from simulations and experiments but also between reflectivity curves and volume fraction profiles [47] by using the NeutronRefTools plugin available in the VMD molecular visualization and analysis software [48]. Information from MD simulations were extracted from both coarse-grained and all-atoms molecular dynamics trajectories and reflectivity and SLD profiles were extracted by slicing the MD trajectory into a discrete number of slabs as shown in Figure 4.

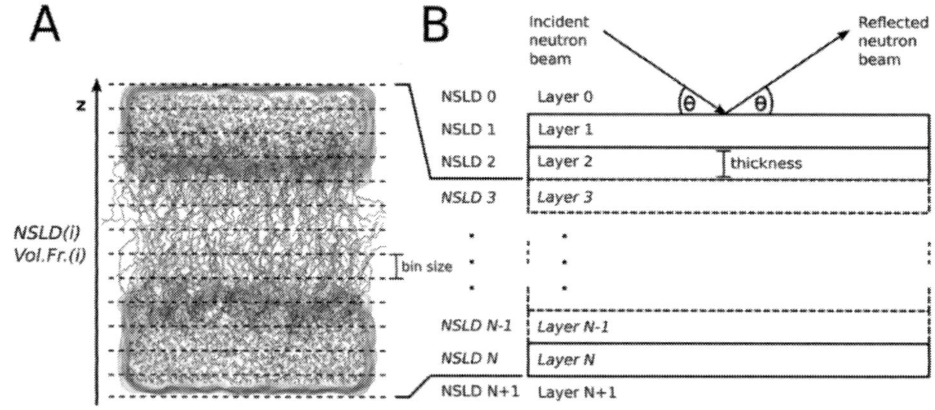

Figure 4. Definition of a slab model for the calculation of reflectivity and SLD profiles from MD trajectories. Reprinted with permission from Ref. [47]. Copyright (2015) American Chemical Society.

4 Examples of applications

This section contains some examples selected from articles already published in literature highlighting the possibilities offered by NR in the field of biological sciences. These examples were chosen to compare the approaches available for the analysis of bio-systems, the variety of samples that can be investigated and the different information that can be obtained by the analysis of NR data. In any case, this represents a very restricted subset of applications of NR to biological systems and most of them were selected for the quality of the work presented and for their pedagogical aspects.

4.1 Biological membranes

Bio-membranes are the molecular assemblies at the basis of life. They are chemically diverse and formidably complex [49–51]. Cell membranes are not only passive barriers that separate the interior of cells from the outside environment protecting the integrity of the cell. They are indeed active barriers able to perform different and fundamental processes such as control of selective exchange of substances between inner and outer cellular environments, regulate cell growth and recognize external stimuli [52]. Because of their nanometric thickness, they are perfect systems to be studied by means of neutron reflectometry [53]. Their study is fundamental for several aspects; they can be used to improve our knowledge on diseases involving cell membranes alterations, to develop new drugs and to understand how vital processes are regulated.

"Simple" lipid bilayers composed by a limited number of synthetic phospholipid species are systems investigated routinely and their structural characterization can be obtained by standard NR measurements. However, these simple systems remain of interest for the investigation of collective fundamental physical properties as fluctuations [54, 55], phase transitions [24, 44] or structural dynamics [17, 56].

In the last ten years, greater interest has been growing towards the investigation of more complex membranes composed by several constituents [57, 58] or obtained from natural systems [59].

Of particular interest is the case of reconstituted Gram-negative bacterial membranes. Gram-negative bacteria are of biomedical and technological interest because of their increasing antibiotic resistance. In Gram-negative bacteria, the outer membrane (OM) acts as a barrier preventing antibiotics to enter in the bacteria and therefore hinders their activity. For this reason, a structural investigation of the OM is of particular interest either in presence or absence of antimicrobial molecules. Considering applications of NR to these systems, Clifton *et al.* [60] fabricated *an accurate in vitro model of the E. coli envelope*. In particular, they

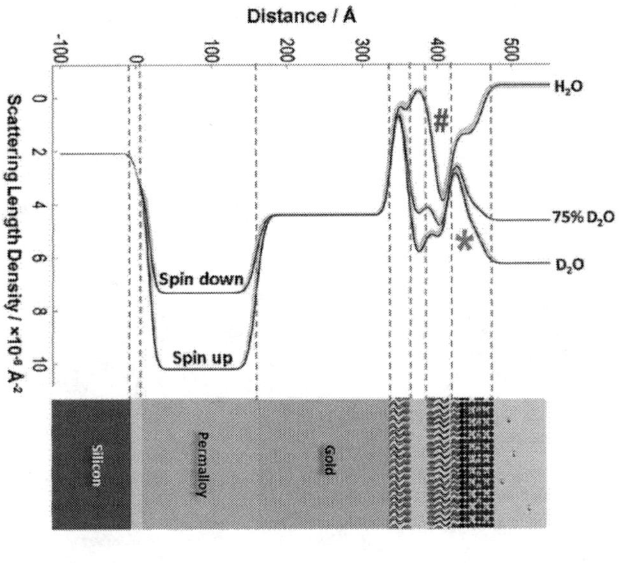

Figure 5. SLD profiles (A) of the OM model in different contrasts together with a schematic representation (B) of the OM model. The permalloy layer was necessary for the use of polarised neutrons. The strong peaks of the deuterated DPPC tails versus H_2O (#) and the non-deuterated LPS tails versus D_2O ($*$), which confirm asymmetry, are evident. Reprinted with permission from Ref. [60].

produced a model system, with a very complex architecture, closely recreating the natural behavior of the *E. coli* OM. The main feature of an OM resides in its asymmetric structure, with lipopolysaccharide (LPS) located in the outer leaflet of the membrane (the one facing the extracellular environment) and phospholipids facing the interior of the cell. This profound asymmetry creates a uniquely impermeable layer that makes OM resistant against the uptake of molecules such as antibiotics [61]. Samples were prepared by Langmuir-Blodgett and Langmuir-Schaefer deposition techniques in order to have a full control on the composition of the inner and outer leaflets of the membrane. The authors deposited the asymmetric OM onto a metallic substrate fractionalised with a self-assembled phosphatidylcholine monolayer. In this way, the OM floats on the top of a water cushion and is decoupled from the solid substrate. To enhance the presence of an asymmetric structure, deuterated phospholipids were used as components on the inner OM leaflet. The authors used neutron reflectometry to reveal, with sub-nanometer resolution, the asymmetric structure of the model OM. The asymmetric structure of the OM mimicking system is shown in Figure 5 in terms of its SLD profiles determined form the analysis of NR data. As reported by the authors, the analysis of such a complex system is however easier than the creation of an accurate *in vitro* systems, which is technically challenging. The structure of the LPS leaflet compared well with the one determined for LPS monolayers at the air-water interface [62]. The structure of the bilayer resulted also compatible with a model generated from atomistic molecular dynamics simulations and reported in [60]. The stability of the model OM was tested under different solution conditions, including presence and absence of divalent cations, EDTA (a chelating agent) and antimicrobial peptides. It is known that *in vivo*, EDTA removes divalent cation and promotes a LPS depletion form the OM, leading to the appearance of phospholipids in the outer leaflet [63]. In their *in vitro* model, Clifton *et al.* observed that bilayer asymmetry was reduced by approximately 20% upon the sequestration of Ca^{2+} by EDTA, confirming thus the possibility to replicate the *in vivo* OM behaviour.

The authors monitored also the effect of antimicrobial proteins, such as lactoferrin, produced by the human immune system. Lactoferrin is a cationic protein and it is supposed to act upon the OM through electrostatic interactions, disrupting divalent-cation bridges between LPS molecules, causing LPS release from the membrane. Since the SLD of lactoferrin differs for that of LPS and phospholipids, it is possible to use NR to localise proteins within the OM structure (see Figure 6). The authors could observed that because of the interaction with

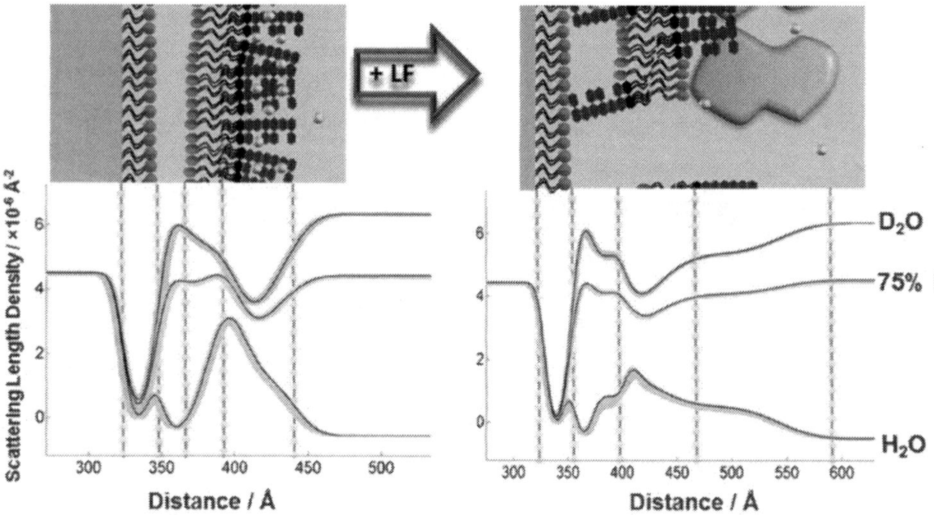

Figure 6. SLD profiles and schematic representation of the OM model before and after addition of lactoferrin. Reprinted with permission from Ref. [60].

lactoferrin, the OM coverage and asymmetry were noticeably reduced. At the same time, the thickness of the LPS core region increased by 90 Å suggesting that proteins were bound to the bilayer with their major axis parallel to the membrane normal. All together, the authors could provide the first structural picture of lactoferrin disrupting the OM.

4.2 Biomolecules at interfaces

Protein adsorption at interfaces represents a very active research area for which neutron reflectometry plays an important role. In fact, while other experimental techniques as QCM-D (Quartz Crystal Microbalance with Dissipation monitoring), ellipsometry, film balance, AFM (Atomic Force Microscopy) and SPR (Surface Plasmon Resonance) can be used to determine the amount of absorbed material and the kinetics of absorption, NR can provide a characterization of the organisation of protein molecules at interfaces. Among many cases, the use of NR is particularly advantageous for the study of i) decorated interfaces promoting protein adsorption at different distances from the interface (for example in presence of polymer brushes) and of ii) adsorption leading to the formation of multilayers.

4.2.1 Proteins at a solid-liquid interface

In the framework of the investigation of protein stability in chromatography processes, Bracewell *et al.* [64] used NR at a solid-liquid interface to study the adsorption of immunoglobulin G (IgG) at various solid-liquid interfaces. We report here the case of adsorption on a pristine silicon-water interface. It is worth recalling that silicon substrates used in NR are typically characterized by the presence of a naturally-growing silica layer on their surface which can be rendered hydrophilic by cleaning or etching processes. The interface of interest is therefore silica and not that of a pure silicon crystals. The surface of the solid substrate was put in contact with an aqueous buffer in which IgG was injected. Measurements were performed using both D_2O and H_2O-based buffers in order to apply the contrast variation method [5]. Data were analysed according to a slab model introduced briefly in Section 3.1. The resulting SLD profiles, together with the relative reflectivity curves are displayed in Figure 7 for the pristine silicon substrate, adsorbed IgG and bound IgG remaining at the interface after a rinsing step. The SLD profiles were interpreted in terms of the amount (volume fraction) of protein present in each of them following the relation

$$\rho_j = \Phi_{w,j}\rho_w + \Phi_{IgG,j}\rho_{IgG} \tag{2}$$

where ρ_j is the total SLD of the j-th layer, ρ_w and ρ_{IgG} the SLD of buffer and of IgG. The volume fractions of buffer and IgG in the j-th layer are $\Phi_{w,j}$ and $\Phi_{IgG,j}$ respectively ($\Phi_{w,j} + \Phi_{IgG,j} \equiv 1$ by definition). Equation 2 is simply the two-components version of Equation 1. By evaluating Φ_{IgG} in each slab and comparing the slabs thickness to the structure of IgG in solution, the authors could conclude that the layer closest to the silica surface represented a population of IgG molecules adsorbed in a "flat on" orientation while the outer layers were ascribed to a sparse self-associating IgG layer in a flat to side-on orientation. The effect of the rinsing step, that caused the removal non-bound IgG molecules, was also visible from the changes in the SLD profiles for large z values.

4.2.2 Antibodies on PEG-brushes

Resistance to protein adsorption onto surfaces is of large interest for biotechnological and pharmaceutical applications. It is known that polyethylene glycol (PEG) chains can resist

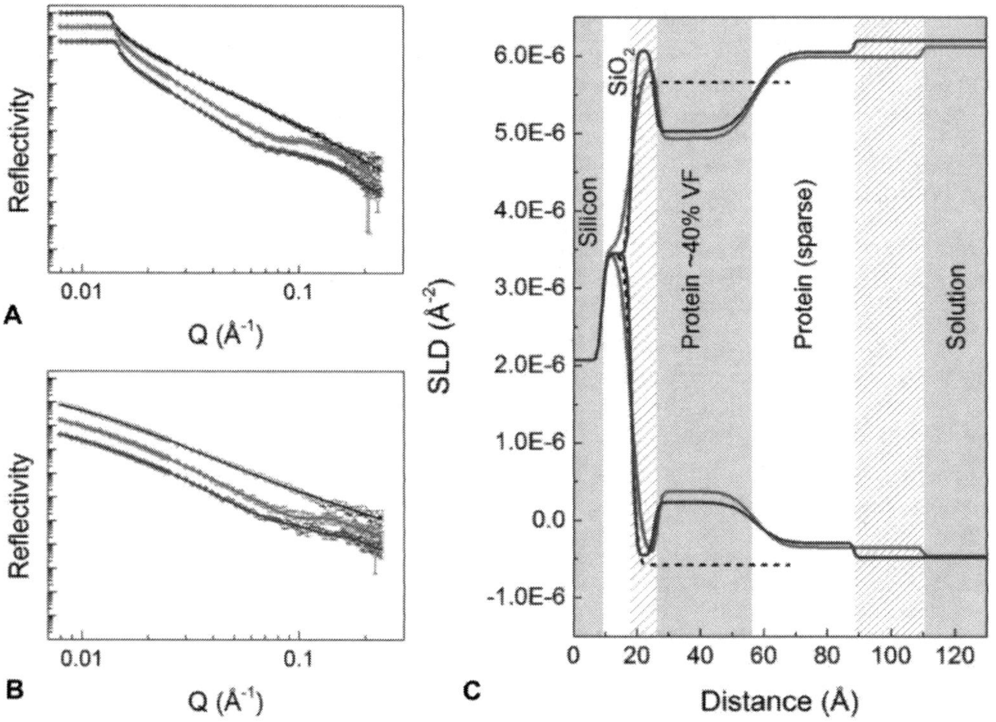

Figure 7. A-B) Reflectivity curves for the pristine silica (black) and for the sequential IgG adsorption (red) and rinse (blue) steps for D_2O (A) and H_2O (B) based buffers. Curves are offset to improve their visibility. C) Scattering length density profiles as a function of the distance z from the substrate surface obtained from the R(Q) curves shown in panels A and B. Reprinted from Ref. [64] under the Creative Commons Attribution License (CC BY).

protein adsorption [65, 66]. In order to understand the molecular mechanisms at the basis of successes and failures of PEG-based surface coatings, Schneck *et al.* [30] used NR to evidence ternary adsorption of antibodies (in particular of anti-PEG Immunoglobulin G) on methoxy terminated PEG brushes. For this purpose, the surface of a silicon crystal was modified by the deposition of an octadecyltrichlorosilane (OTS) layer onto which lipid monolayers anchoring the PEG brushes were deposited by the Langmuir-Schaefer technique. The surface was exposed to a water-based buffer in which IgG antibodies (Ab) were dispersed. NR measurements were performed after incubation and rinsing steps and collected in different H/D contrasts. NR data were then analysed using a component groups approach similar to the one described in Section 3.2. Figure 8 shows the resulting volume fraction distributions together with a pictorial sketch describing the structure of the investigated system. NR experiments were performed for different brushes grafting density and the authors could evidence a clear increase of the adsorbed amount of Ab as the grafting density increased. From the analysis of the volume fraction profiles, the authors could determine the type of Ab adsorption, concluding that the latter took place within the PEG brush because of an attractive interaction between Abs and PEG (ternary adsorption). Moreover, a more detailed interpretation of the volume fraction profiles indicated that the antibodies adsorbed in an inverted "Y" configuration, interacting mainly with the terminal part of the PEG chains.

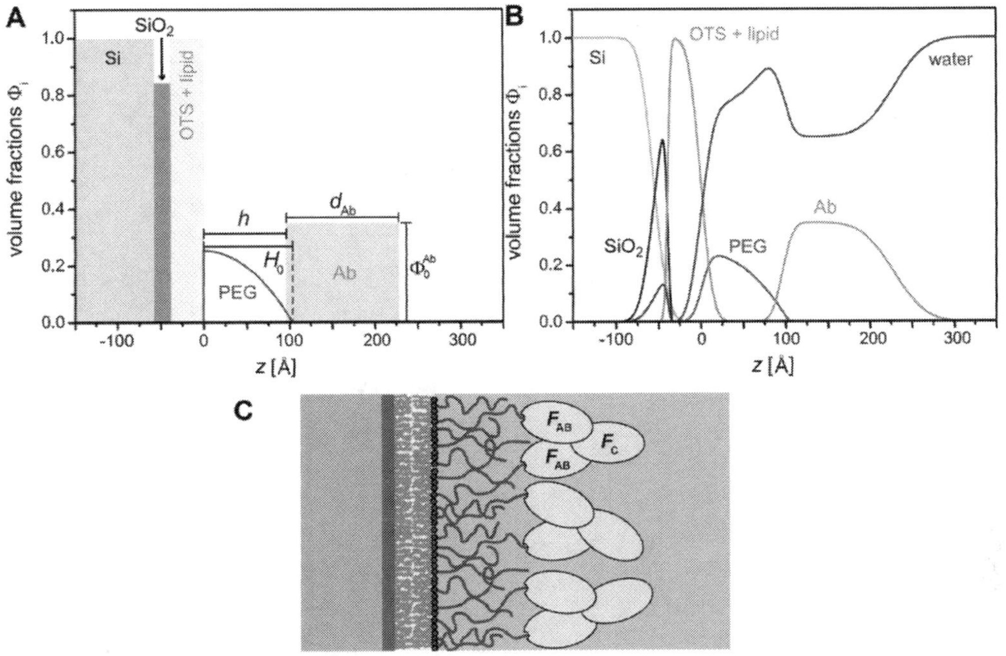

Figure 8. A) Idealised volume fraction model before optimisation, B) optimised volume fraction distribution obtained from the analysis of NR data and C) pictorial representation of the equilibrated structure showing Ab binding. Reproduced from Ref. [30] with permission from Elsevier.

4.3 Interactions with membranes

Interactions of small natural and synthetic molecules with lipids permeate our daily life. At the level of the cell membrane, these interactions are essential for the proper functioning of living organisms and play important roles in diseases, in their diagnosis and in therapeutic processes. They are also or primary interest in food sciences, being lipid molecules among the most common components of food. Lipid monolayers can be found as stabilisers at many interfaces and in emulsions and therefore they have a major role in food preservation, food digestion and fat uptake. Among all these several possible examples, this section will be focused on two important studies reporting on the cytotoxicity of amyloid polypeptide and on the antimicrobial action of a well known drug, Amphotericin B.

4.3.1 The case of an amyloid polypeptide

As described in Sections 4.1 and 4.2, lipids and proteins films represent typical samples investigated by NR. Increasing attention has been devoted, in the last 10 years, to the study of the interactions between these two biomolecules. In fact, protein-membrane interactions are of general great interest; they are at the basis of many diseases (including some of high societal impact as Alzheimer's and Parkinson's) and, at the same time, they can be exploited for developing novel antimicrobial and therapeutic strategies. Peptide–membrane interaction mechanisms have been the subject of numerous studies and debates in the past decades and are still a puzzle. Neutrons scattering techniques, and in particular neutron reflectometry, have been used in all these contexts. Among several possible examples, the case of amyloid peptides involved in neurodegenerative and systemic diseases [67] was chosen. Martel *et al.*

[68] used NR, together with other experimental techniques, to investigate the amyloidogenicity and cytotoxic mechanisms of an amyloid polypeptide (IAPP) involved in type 2 Diabetes Mellitus. In order to identify which peptide fragments (genes) are promoting interactions with membranes (and therefore involved in cytotoxic effects related to the type 2 diabetes Mellitus), structural changes induced on model-membranes by the human peptide were compared to those induced by rat and mutagen peptides. This comparison was performed since rats are not affected by this disease even if they host an amyloid polypeptide characterized by a genetic sequence very similar to that of the human one. By performing NR experiments, the author could quantify and localise the structural changes induced in a solid supported bilayer (SLB) as reported in Figure 9. The SLD profiles shown in Figure 9 refer to the pep-

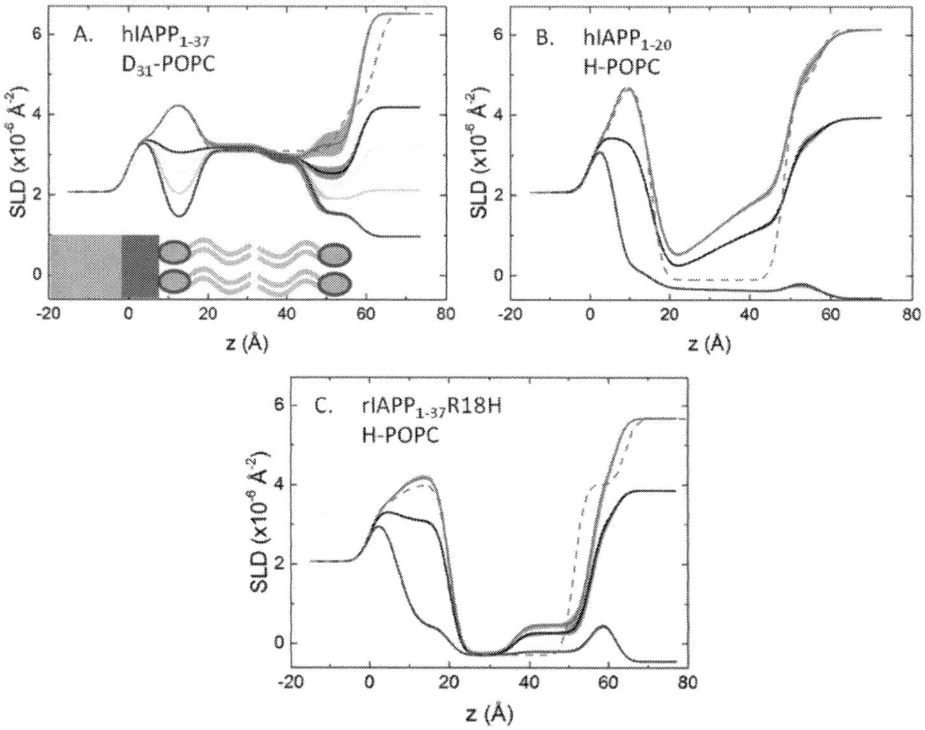

Figure 9. SLD profiles obtained from the analysis of NR curves recorded for SLBs (A d_{31}POPC, B-C POPC) in presence of different peptides and at different contrasts: 0% D_2O in blue, 38% D_2O in green, 52% D_2O in yellow, 66% D_2O in black, 100% D_2O in red. The SLD curves of pristine SLBs in 100% D_2O buffer are also shown as dotted lines. Reprinted with permission from Ref. [68]. Copyright (2016) American Chemical Society.

tides that were found to induce alterations in the bilayer structure. They indicated that the full length human IAPP (hIAPP$_{1-37}$ in panel A in Figure 9) interacted only with the outer leaflet (see Figure 1) of the SLB. The change in the profiles was interpreted as a removal of phospholipid molecules. A similar, but larger, effect was observed for the mutagen rat peptide rIAPP$_{1-37}$R18H (panel C in Figure 9), in which the residue 18 of the original rat peptide was exchanged for the human one. This simple modification induced an important peptide-membrane interaction, otherwise absent for the full length rat peptide [68]. The largest effect, indicating the largest cytotoxicity, was observed for the fragment corresponding to the first 20 residues of the human IAPP (hIAPP$_{1-20}$ in panel B in Figure 9). In this case the membrane was partially destroyed as indicated by the large split, in the z-range corresponding to the hydrocarbon region, between the SLD profiles measured in different contrast conditions for the same sample. Such a split indicated water permeation of the SLB and it was related to the formation of large, macroscopic pores. By comparing these results with those obtained from other experimental techniques and simulations, the authors described a new scenario in which the peptide toxicity is not caused by the formation of amyloid fibers (on the contrary to other famous disease as Alzheimer's) but by the effect of a membrane-disturbing peptide oligomer, that lead to the formation of pores in the membrane. For this class of peptides, already existing hypotheses on the link between the formation of large amyloid aggregates and self-protection mechanisms inhibiting peptide toxicity were confirmed. Indeed, the toxic portions of the peptide were isolated within the aggregates and their interaction with cell membranes was, in this way, avoided.

4.4 The case of Amphotericin B

Amphotericin B (AmB) is a highly toxic antifungal drug used to treat systemic fungal infections. It is now widely accepted that AmB action is mediated by its interaction with the ergosterol present in yeast and fungal membranes [69]. However, the molecular mechanisms at the base of this interaction are still object of a large debate. de Ghellinck *et al.* [59] investigated the activity of AmB on SLBs composed by natural lipids derived from *Pichia Pastoris* yeast [6], with a particular focus on the role of the native sterol components present in the SLBs. From the lipid analysis, ergosterol resulted to be the predominant sterol molecule in the *Pichia Pastoris* yeast extracts. NR data, collected in different contrast conditions and in presence or absence of sterols, were analysed according to the Abelès' matrix formalism [21] and the results of the analysis were expressed in terms of SLD profiles. Upon addition of AmB changes in these profiles were observed and their interpretation lead to the understanding of the action mechanism of AmB. In fact, in absence of sterols the changes indicated the insertion of AmB in the membrane as well as an accumulation of AmB molecules on the membrane surface (see Figures 10a-b and e-f). In presence of sterol (Figures 10c-d and g-h), the latter were removed from the membrane upon addition of AmB; the SLB final structure contained no sterol and a low amount of AmB, while accumulation of AmB and sterols on the surface of the membrane was detected. These result represented the first structural investigation of AmB effects on yeast lipid membranes. The membranes reconstituted from natural lipid extracts reported in the work of de Ghellinck *et al.* [59] had a large degree of poly-unsaturation, that resulted crucial for the AmB interaction. In fact, the authors observed that AmB inserted into membranes even in the absence of sterols and that the inserted amount and the effect on the membrane structure critically depended on the lipid chain composition. The most important result they obtained was about the sterol depletion caused by AmB. In fact, AmB resulted to extract all the sterols from the SLBs, transferring them to a diffuse layer carpeting the surface of the membrane. This sterol depletion induced a significant membrane thinning that the authors related to the drug cytotoxicity. In fact, such a pronounced membrane thinning is likely to interfere with a broad range of membrane functions.

5 Conclusions

The collection of results reported in this chapter aimed to highlight the possibilities that an advanced scattering technique, as neutron reflectometry, can have for biologically relevant systems. As described in the previous sections, the data analysis approaches evolved to respond to the increase in samples complexity. Among the most relevant biological samples that can be investigated by NR, lipid membranes will continue to represent one of the most

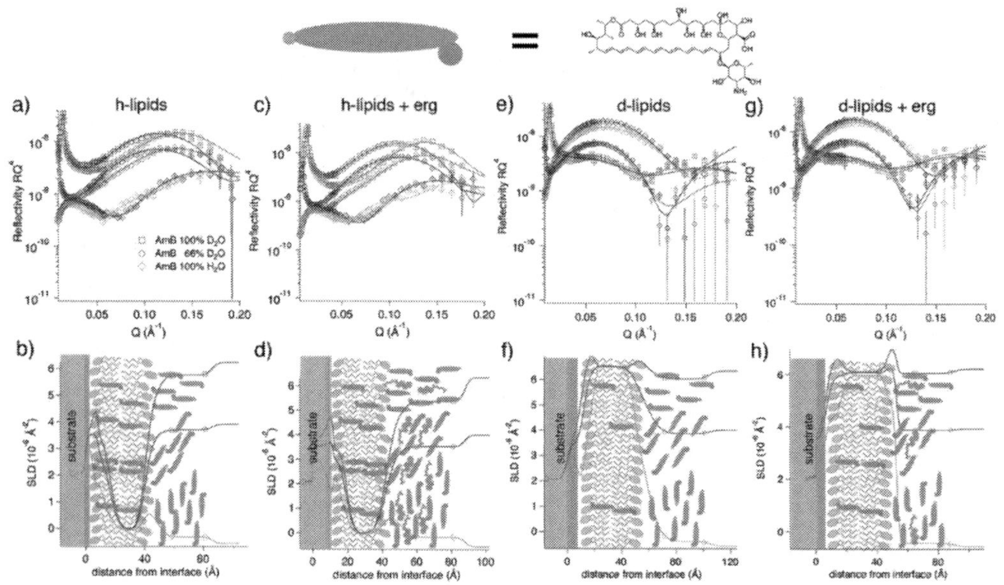

Figure 10. Reflectivity curves and corresponding SLD profiles illustrating the effect of AmB on SLBs composed *P. pastoris* lipid extracts in absence (a-b, e-f) and in presence (c-d, g-h) of the native sterols upon addition of 1 mM AmB in different contrast conditions (different H_2O:D_2O ratio). The cartoons show the possible location of AmB corresponding to the layer thicknesses in the structures formed. Reprinted from Ref. [59], Copyright (2015), with permission from Elsevier.

relevant topics, because of their interest and implications in many aspects of our daily life. At the same time study of thin films in the context of novel bio-materials is expected to increase noticeably. In both cases, the use of natural biomolecules is expected to play a significant role. For this reason, new developments will be focused not only on the technique itself but mostly on sample production and preparation approaches. At the same time, the investigation of model systems will remain a hot topic in the physical chemistry and biophysics communities since NR allows to measure fundamental collective properties not accessible by means of other techniques.

References

[1] Z. Fisher, A. Jackson, A. Kovalevsky, E. Oksanen, H. Wacklin, in *Exp. Methods Phys. Sci.* (Academic Press, 2017), Vol. 49, pp. 1–75, ISBN 9780128053249

[2] M. Vaccaro, G. Mangiapia, L. Paduano, E. Gianolio, A. Accardo, D. Tesauro, G. Morelli, ChemPhysChem **8**, 2526 (2007)

[3] R. Pignatello, T. Musumeci, L. Basile, C. Carbone, G. Puglisi, J. Pharm. Bioallied Sci. **3**, 4 (2011)

[4] S. Mitragotri, D.G. Anderson, X. Chen, E.K. Chow, D. Ho, A.V. Kabanov, J.M. Karp, K. Kataoka, C.A. Mirkin, S.H. Petrosko et al., ACS Nano **9**, 6644 (2015)

[5] T.L. Crowley, E.M. Lee, E.A. Simister, R.K. Thomas, Phys. B Phys. Condens. Matter **173**, 143 (1991)

[6] A. de Ghellinck, H. Schaller, V. Laux, M. Haertlein, M. Sferrazza, E. Maréchal, H. Wacklin, J. Jouhet, G. Fragneto, PLoS One **9**, e92999 (2014)

[7] M. Moulin, G.A. Strohmeier, M. Hirz, K.C. Thompson, A.R. Rennie, R.A. Campbell, H. Pichler, S. Maric, V.T. Forsyth, M. Haertlein, Chem. Phys. Lipids **212**, 80 (2018)
[8] S. Waldie, M. Moulin, L. Porcar, H. Pichler, G.A. Strohmeier, M. Skoda, V.T. Forsyth, M. Haertlein, S. Maric, M. Cárdenas, Sci. Rep. **9**, 5118 (2019)
[9] M. Haertlein, M. Moulin, J.M. Devos, V. Laux, O. Dunne, V. Trevor Forsyth, in *Methods Enzymol.* (Academic Press, 2016), Vol. 566, pp. 113–157, ISBN 9780128030493
[10] M.C. Petty, W.A. Barlow, *Film Deposition* (Springer US, Boston, MA, 1990), pp. 93–132, ISBN 978-1-4899-3716-2
[11] A.P. Girard-Egrot, L.J. Blum, in *Nanobiotechnology Biomim. Membr.* (Springer US, Boston, MA, 2007), pp. 23–74
[12] H.P. Wacklin, Langmuir **27**, 7698 (2011)
[13] H. Mohwald, Annu. Rev. Phys. Chem. **41**, 441 (1990)
[14] C. Schladitz, E.P. Vieira, H. Hermel, H. Möhwald, Biophys. J. **77**, 3305 (1999)
[15] U. Mennicke, T. Salditt, Langmuir **18**, 8172 (2002)
[16] S. Micciulla, D.W. Hayward, Y. Gerelli, A. Panzarella, R. von Klitzing, M. Gradzielski, L. Chiappisi, Commun. Chem. **2**, 61 (2019)
[17] Y. Gerelli, L. Porcar, G. Fragneto, Langmuir **28**, 15922 (2012)
[18] V. Rondelli, P. Brocca, N. Tranquilli, G. Fragneto, E. Del Favero, L. Cantù, Biophys. Chem. **229**, 135 (2017)
[19] G. Fragneto, T. Charitat, J. Daillant, Eur. Biophys. J. **41**, 863 (2012)
[20] V. Rondelli, G. Fragneto, S. Motta, E.D. Favero, L. Cantù, J. Phys. Conf. Ser. **340**, 012083 (2012)
[21] F. Abelès, Ann. Phys. (Paris). **12**, 706 (1950)
[22] L.G. Parratt, Phys. Rev. **95**, 359 (1954)
[23] R.A. Campbell, Y. Saaka, Y. Shao, Y. Gerelli, R. Cubitt, E. Nazaruk, D. Matyszewska, M.J. Lawrence, J. Colloid Interface Sci. **531**, 98 (2018)
[24] Y. Gerelli, Phys. Rev. Lett. **122**, 248101 (2019)
[25] Y. Gerelli, J. Appl. Crystallogr. **49**, 330 (2016)
[26] A. Nelson, J. Appl. Crystallogr. **39**, 273 (2006)
[27] M. Björck, G. Andersson, J. Appl. Crystallogr. **40**, 1174 (2007), 1503.07163
[28] A.V. Hughes, F. Ciesielski, A.C. Kalli, L.A. Clifton, T.R. Charlton, M.S.P. Sansom, J.R.P. Webster, Acta Crystallogr. Sect. D Struct. Biol. **72**, 1227 (2016)
[29] A.R. Nelson, S.W. Prescott, J. Appl. Crystallogr. **52**, 193 (2019)
[30] E. Schneck, I. Berts, A. Halperin, J. Daillant, G. Fragneto, Biomaterials **46**, 95 (2015)
[31] V.M. Latza, I. Rodriguez-Loureiro, I. Kiesel, A. Halperin, G. Fragneto, E. Schneck, Langmuir **33**, 12708 (2017)
[32] S. Micciulla, Y. Gerelli, R.A. Campbell, E. Schneck, Langmuir **34**, 789 (2018)
[33] S. Micciulla, Y. Gerelli, E. Schneck, Biophys. J. **116**, 1259 (2019)
[34] M. Wiener, S. White, Biophys. J. **61**, 434 (1992)
[35] N. Kučerka, J.F. Nagle, J.N. Sachs, S.E. Feller, J. Pencer, A. Jackson, J. Katsaras, Biophys. J. **95**, 2356 (2008)
[36] M. Belička, Y. Gerelli, N. Kučerka, G. Fragneto, Soft Matter **11**, 6275 (2015)
[37] J. Pan, X. Cheng, F.A. Heberle, B. Mostofian, N. Kučerka, P. Drazba, J. Katsaras, J. Phys. Chem. B **116**, 14829 (2012)
[38] N. Kučerka, B.W. Holland, C.G. Gray, B. Tomberli, J. Katsaras, J. Phys. Chem. B **116**, 232 (2012)
[39] L. Tavagnacco, Y. Gerelli, A. Cesàro, J.W. Brady, J. Phys. Chem. B **120**, 9987 (2016)

[40] P.c. Chen, R. Shevchuk, F.M. Strnad, C. Lorenz, L. Karge, R. Gilles, A.M. Stadler, J. Hennig, J.S. Hub, J. Chem. Theory Comput. **15**, 4687 (2019)

[41] A.A. Mungikar, D. Forciniti, *Computer simulations and neutron reflectivity of proteins at interfaces* (2002)

[42] C.F. Majkrzak, N.F. Berk, S. Krueger, J.A. Dura, M. Tarek, D. Tobias, V. Silin, C.W. Meuse, J. Woodward, A.L. Plant, Biophys. J. **79**, 3330 (2000)

[43] S. Bobone, Y. Gerelli, M. De Zotti, G. Bocchinfuso, A. Farrotti, B. Orioni, F. Sebastiani, E. Latter, J. Penfold, R. Senesi et al., Biochim. Biophys. Acta - Biomembr. **1828**, 1013 (2013)

[44] A. Koutsioubas, J. Phys. Chem. B **120**, 11474 (2016)

[45] F. Heinrich, M. Lösche, Biochim. Biophys. Acta - Biomembr. **1838**, 2341 (2014)

[46] A.R. McCluskey, J. Grant, A.J. Smith, J.L. Rawle, D.J. Barlow, M.J. Lawrence, S.C. Parker, K.J. Edler, J. Phys. Commun. **3**, 075001 (2019)

[47] L. Darré, J. Iglesias-Fernandez, A. Kohlmeyer, H. Wacklin, C. Domene, J. Chem. Theory Comput. **11**, 4875 (2015)

[48] W. Humphrey, A. Dalke, K. Schulten, J. Mol. Graph. **14**, 33 (1996)

[49] K. Simons, E. Ikonen, Nature **387**, 569 (1997)

[50] K. Simons, W.L. Vaz, Annu. Rev. Biophys. Biomol. Struct. **33**, 269 (2004)

[51] K. Simons, M.J. Gerl, Nat. Rev. Mol. Cell Biol. **11**, 688 (2010)

[52] P.L. Yeagle, FASEB J. **3**, 1833 (1989)

[53] G. Fragneto, Eur. Phys. J. Spec. Top. **213**, 327 (2012)

[54] T. Salditt, C. Münster, U. Mennicke, C. Ollinger, G. Fragneto, Langmuir **19**, 7703 (2003)

[55] J. Daillant, E. Bellet-Amalric, A. Braslau, T. Charitat, G. Fragneto, F. Graner, S. Mora, F. Rieutord, B. Stidder, Proc. Natl. Acad. Sci. U. S. A. **102**, 11639 (2005)

[56] Y. Gerelli, L. Porcar, L. Lombardi, G. Fragneto, Langmuir **29**, 12762 (2013)

[57] V. Rondelli, E. Del Favero, S. Motta, L. Cantù, G. Fragneto, P. Brocca, Eur. Phys. J. E **36**, 73 (2013)

[58] G. Fragneto, R. Delhom, L. Joly, E. Scoppola, Curr. Opin. Colloid Interface Sci. **38**, 108 (2018)

[59] A. de Ghellinck, G. Fragneto, V. Laux, M. Haertlein, J. Jouhet, M. Sferrazza, H. Wacklin, Biochim. Biophys. Acta - Biomembr. **1848**, 2317 (2015)

[60] L.A. Clifton, S.A. Holt, A.V. Hughes, E.L. Daulton, W. Arunmanee, F. Heinrich, S. Khalid, D. Jefferies, T.R. Charlton, J.R. Webster et al., Angew. Chemie - Int. Ed. **54**, 11952 (2015)

[61] J.M. Pagès, C.E. James, M. Winterhalter, *The porin and the permeating antibiotic: A selective diffusion barrier in Gram-negative bacteria* (2008)

[62] A.P. Le Brun, L.A. Clifton, C.E. Halbert, B. Lin, M. Meron, P.J. Holden, J.H. Lakey, S.A. Holt, Biomacromolecules **14**, 2014 (2013)

[63] M. Vaara, Microbiol. Rev. **56**, 395 (1992)

[64] A.R. Mazzer, L.A. Clifton, T. Perevozchikova, P.D. Butler, C.J. Roberts, D.G. Bracewell, J. Chromatogr. A **1499**, 118 (2017)

[65] J.M. Harris, ed., *Poly(Ethylene Glycol) Chemistry* (Springer US, Boston, MA, 1992), ISBN 978-1-4899-0705-9

[66] H. Chen, L. Yuan, W. Song, Z. Wu, D. Li, Prog. Polym. Sci. **33**, 1059 (2008)

[67] R.S. Harrison, P.C. Sharpe, Y. Singh, D.P. Fairlie, in *Rev. Physiol. Biochem. Pharmacol.* (Springer Berlin Heidelberg, Berlin, Heidelberg, 2007), Vol. 159, pp. 1–77

[68] A. Martel, L. Antony, Y. Gerelli, L. Porcar, A. Fluitt, K. Hoffmann, I. Kiesel, M. Vivaudou, G. Fragneto, J.J. de Pablo, J. Am. Chem. Soc. **139**, 137 (2017)
[69] K.C. Gray, D.S. Palacios, I. Dailey, M.M. Endo, B.E. Uno, B.C. Wilcock, M.D. Burke, Proc. Natl. Acad. Sci. **109**, 2234 (2012)

EPJ Web of Conferences **236**, 05001 (2020) https://doi.org/10.1051/epjconf/202023605001
JDN 24

Probing the dynamics of biological matter by elastic, quasi-elastic, and inelastic neutron scattering

Giorgio Schirò[1,*]

[1] Université Grenoble Alpes, CEA, CNRS, Institut de Biologie Structurale, F-38000, Grenoble, France

Abstract. The so-called function-structure-dynamics paradigm established that a close relationship links the way biological molecules work (function), their 3-dimensional organization (structure) and the changes of this organization in time (dynamics), which characterize biomolecules as highly dynamic objects. A typical example of protein dynamics is provided by protein reactions with substrates: equilibrium thermal fluctuations of protein structure are necessary to allow the access of substrates to the active site, where the functional reaction occurs. Neutron scattering is a powerful technique to study equilibrium protein structural dynamics. The incoherent structure factor, which is dominant in neutron scattering from biological matter, is related to the time-position self correlation function of protein/solvent nuclei. Here the basic theory of neutron scattering and the principles of the technologies used to measure it are described. Some selected applications of neutron scattering for investigating the structural dynamics of biological molecules are also reviewed.

1 Theoretical background

This section contains the basic formalism for describing neutron scattering in a quantitative way and forms the basis for understanding the following sections dedicated to neutron scattering from biological matter. We will show how the macroscopic physical quantities measured by neutron scattering are related to the microscopic state of the matter interacting with neutrons during an experiment. More detailed descriptions of the theory of neutron scattering can be found in the textbooks that have been used as the basis for the short description reported here [1–3].

The neutron is an electrically neutral nuclear particle with a mass $m_n = 1.675 \times 10^{-27}$ kg and a spin $s = 1/2$. The neutron does not live naturally in free form, but decays into a proton, an electron, and an antineutrino, with a lifetime $\tau \sim 886$ s, much longer than the time of a scattering event during an experiment, typically a fraction of a second. The neutron interacts with nuclei via the strong nuclear force and with magnetic moments via dipole-dipole coupling.

*Corresponding author: giorgio.schiro@ibs.fr

1.1 Scattering cross sections

We first introduce the terms used to describe the scattering of a neutron beam. We define the flux of a neutron beam as

$$\Psi = \frac{number\ of\ neutrons\ incident\ on\ a\ surface\ per\ second}{surface\ area\ perpendicular\ to\ the\ neutron\ beam\ direction} \tag{1}$$

The central parameter to describe the interaction of a neutron beam with matter is the neutron scattering cross section σ defined as

$$\sigma = \frac{1}{\Psi} \cdot number\ of\ neutrons\ scattered\ per\ second \tag{2}$$

which has the units of an area. The angular dependence of the scattered neutrons is the most important aspect of the scattering process. To describe this we use the so-called differential scattering cross section, defined as:

$$\frac{d\sigma}{d\Omega} = \frac{1}{\Psi} \cdot \frac{number\ of\ neutrons\ scattered\ per\ second\ in\ the\ solid\ angle\ d\Omega}{d\Omega} \tag{3}$$

We consider now the possibility that, during the scattering process, incident neutrons exchange part of their energy with the target nuclei, which is defined as inelastic scattering. Conversely, if incident neutrons are scattered with no energy exchange we refer to as elastic scattering. In order to describe the inelastic scattering process we have to take into account the energy dependence of the scattered neutrons. We define the double differential scattering cross section as:

$$\frac{d^2\sigma}{d\Omega dE} = \frac{1}{\Psi} \cdot \frac{n.\ of\ neutrons\ scattered\ per\ second\ in\ solid\ angle\ d\Omega\ with\ energies\ [E;E+dE]}{d\Omega} \tag{4}$$

In the neutron scattering experiments that will be described here, only the strong nuclear interaction between neutrons and nuclei in the sample is considered. We define the state of the incoming wave as:

$$|\psi_i> \propto exp\,(i\,\widehat{k_0} \cdot \widehat{r}) \tag{5}$$

where $\widehat{k_0}$ is the wave vector of the incoming neutron, $\widehat{r}$ is its position vector and the symbol $\propto$ indicates that the normalization factor has been omitted. We express the final neutron state at the detector as (a superposition of) plane waves, each with the form:

$$|\psi> \propto exp\,(i\,\widehat{k} \cdot \widehat{r}) \tag{6}$$

We describe the interaction responsible for the scattering by the operator $\hat{V}$. By using the Fermi Golden Rule, which defines the rate of change between the neutron in the incoming state and that in the final state (we omit here some straightforward derivations, in particular the calculation of the density of states), we can obtain an expression for the differential scattering cross section:

$$\frac{d\sigma}{d\Omega} \propto \frac{\hat{k}}{\hat{k}_0}\left(\frac{m_n}{2\pi\hbar^2}\right)^2 \left|<\psi_0|\hat{V}|\psi>\right|^2 \tag{7}$$

To calculate the cross section we need an explicit form for the interaction operator $\hat{V}$. The interaction between the neutron and the nuclei can be approximated by the Fermi pseudopotential defined as:

$$\hat{V} = \frac{2\pi\hbar^2}{m_n} b_N \delta(\hat{r} - \hat{r}_N) \tag{8}$$

where $\hat{r}_N$ is the position vector of the target nucleus in the laboratory frame. Here the main parameter is the scattering length b_N. The spatial δ function accounts for the short range of the strong nuclear forces and gives a sufficient description of scattering of thermal neutrons. For a system composed by a single nucleus, we can now calculate the scattering cross section by using Eq. (7). We start by calculating the matrix element:

$$\left|<\psi_0|\hat{V}|\psi>\right| = \frac{2\pi\hbar^2}{m_n} b_N \int exp\,(i\,\hat{k}_0 \cdot \hat{r})\,\delta(\hat{r} - \hat{r}_N)\,exp(-i\,\hat{k} \cdot \hat{r})\,d^3r =$$

$$= \frac{2\pi\hbar^2}{m_n} b_N exp(i\hat{Q} \cdot \hat{r}_N) \tag{9}$$

where $\hat{Q} = \hat{k}_0 - \hat{k}$ is the momentum transfer. Inserting Eq. (9) into Eq. (7), we obtain:

$$\frac{d\sigma}{d\Omega} = b_N^2\,. \tag{10}$$

As already mentioned, during the scattering event generally neutrons can transfer energy to or absorb energy from the scattering system. We indicate the neutron energy transfer by:

$$\hbar\omega = E_i - E\ \frac{\hbar^2(k_0^2 - k^2)}{2m_n}\,. \tag{11}$$

In this case, Eq. (7) assumes the form:

$$\frac{d\sigma}{d\Omega} \propto \frac{\hat{k}}{\hat{k}_0}\left(\frac{m_n}{2\pi\hbar^2}\right)^2 \left|< \psi_0, E_0|\hat{V}|\psi, E >\right|^2 \delta(E - E_0 - \hbar\omega) \tag{12}$$

where the δ function expresses the energy conservation in the scattering event. In a system of N nuclei, possibly with different scattering lengths, interferences between scattered waves from different nuclei have to be considered. We will neglect here absorption and multiple scattering effects, that can actually be minimized in certain experimental conditions, but also quantitatively taken into account in case such conditions cannot be realized [4]. The differential cross section reads as:

$$\frac{d^2\sigma}{d\Omega d\omega} \propto \frac{\hat{k}}{\hat{k}_0} \sum_{i,j} b_i b_j \frac{1}{2\pi} \int_{-\infty}^{+\infty} e^{i\omega t} \langle exp(-i\hat{Q}\cdot\hat{r}_i(0))\, exp(i\hat{Q}\cdot\hat{r}_j(t))\rangle dt \tag{13}$$

where the indexes i and j are over all nuclei, $\hat{r}_i(t)$ and $\hat{r}_j(t)$ denote the position of nuclei i and j at time t, and $\langle\ \rangle$ indicates the ensemble average. We have also assumed here that the system is stationary.

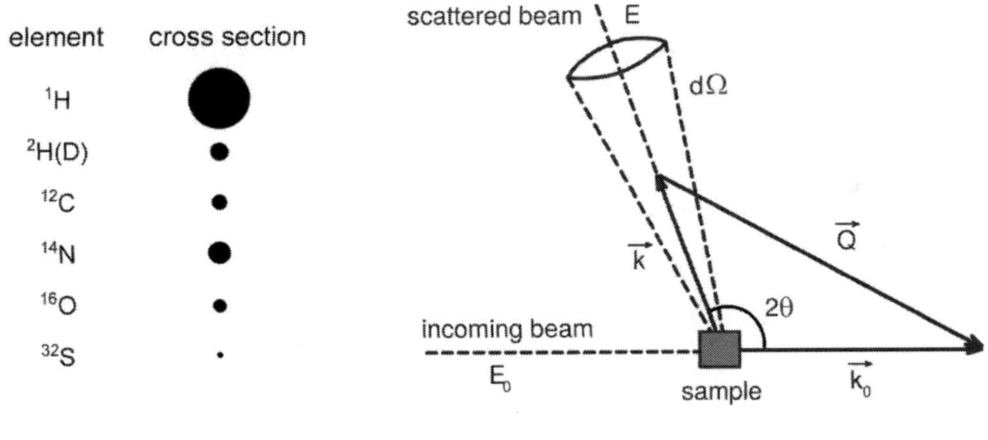

Fig. 1. Left: neutron scattering total cross sections of the most abundant elements present in biological molecules (the area of black circles is proportional to the total cross sections). Right: scheme of the scattering process. An incoming neutron with a wave vector $\hat{k}_0$ interacts with nuclei in the sample and is scattered with a wave vector $\hat{k}$. The fundamental parameters of the process are the momentum transfer $\hat{Q} = \hat{k}_0 - \hat{k}$ and the exchanged energy E - E_0.

In general, the neutron scattering length b varies, the variation being due to the heterogeneity originating from isotopes of the same element, or from different elements. This represents a static disorder that, in the case of a macroscopic sample, can be treated in the same way as a time-dependent disorder (which corresponds to the ergodic hypothesis), since we can assume that the sample is large enough to essentially represent an ensemble average. We assume that the scattering length at site i has the stochastic value $b = \bar{b} \pm \delta b$, where $\bar{b}$ is shorthand for the average of b and the deviation from the average is assumed independent from specific site and is defined as $\delta b = \sqrt{\overline{b^2} - \overline{b}^2}$. Generally, the explicit average notation $\overline{b}$ drops, and the symbol b is used to indicate the average scattering length of a given isotope or element. The total cross section defined in Eq. (13) can then be decomposed into two different terms; one is called coherent and the other one incoherent and are defined as:

$$\frac{d^2\sigma}{d\Omega d\omega} = \left(\frac{d^2\sigma}{d\Omega d\omega}\right)_{inc} + \left(\frac{d^2\sigma}{d\Omega d\omega}\right)_{coh} \propto$$

$$\propto \frac{\hat{k}}{\hat{k}_0}(\overline{b^2} - \overline{b}^2)\sum_i \frac{1}{2\pi}\int_{-\infty}^{+\infty} e^{i\omega t}\langle exp(-i\hat{Q}\cdot\hat{r}_i(0))\, exp(i\hat{Q}\cdot\hat{r}_i(t))\rangle dt +$$

$$+\frac{\hat{k}}{\hat{k}_0}\overline{b}^2\sum_{i,j}\frac{1}{2\pi}\int_{-\infty}^{+\infty} e^{i\omega t}\langle exp(-i\hat{Q}\cdot\hat{r}_i(0))\, exp(i\hat{Q}\cdot\hat{r}_j(t))\rangle dt =$$

$$= \frac{\hat{k}}{\hat{k}_0}(\overline{b^2} - \overline{b}^2)S_{inc}(\hat{Q},\omega) + \frac{\hat{k}}{\hat{k}_0}\overline{b}^2 S_{coh}(\hat{Q},\omega)\ . \qquad (14)$$

The terms $S_{inc}(Q,\omega)$ and $S_{coh}(Q,\omega)$ are known as incoherent and coherent dynamic structure factors, respectively. $S_{coh}(Q,\omega)$, which results from the coherent superposition of the scattering from pairs of scatterers, accounts for the correlation among positions of different nuclei, whereas $S_{inc}(Q,\omega)$, which does not contain any phase information, accounts for the correlation among positions of a single nucleus at different times. The difference between coherent and incoherent scattering can be easily explained if we consider the elastic scattering from a crystal, i.e. from atoms not moving. The elastic coherent scattering as a function of the scattering angle leads to Bragg peaks containing information about the relative position of the scattering nuclei. Conversely, the elastic signal originating from incoherent scattering is flat and the only information we can extract from it is the number of scatterers. We consider now a system composed by moving atoms. In this case the wavefronts emitted at different times from the same particle are coherent and lead to interference among the different scattered waves. The incoherent signal then provides information on the single particle motions. Incoherent and coherent cross sections are usually defined as:

$$\sigma_{inc} = 4\pi\,(\overline{b^2} - \overline{b}^2)$$

$$\sigma_{coh} = 4\pi\,\overline{b}^2\ . \qquad (15)$$

The scattering cross sections vary in an irregular manner as a function of element type and isotope mass number. The case of the hydrogen atom is particularly interesting for the study of soft matter and, in particular, of biological matter. Its nucleus is composed of a single proton, with spin ½. The incoherent scattering cross section of the hydrogen atom is much larger than the incoherent or coherent scattering cross sections of all other elements. This characteristic makes incoherent neutron scattering an ideal tool for the study of individual motions in materials abundant in hydrogen. In the case of deuterium (^2H or D), the relative part of incoherent scattering over the total scattering is strongly reduced with respect to hydrogen ^1H.

In view of both the natural abundance of hydrogen and its large incoherent cross section, if compared with other elements in biological molecules (see Fig. 1), it is evident why neutron scattering represents a technique ideal to study the dynamics of biological systems. In particular, as hydrogens are abundant and uniformly distributed in proteins, this technique gives average information on the dynamics of a protein, as we will see in the next sections.

1.2 Scattering functions

We now want to relate the observables describing the neutron scattering event, as introduced in the previous section, with the microscopic properties of the matter that interacts with the neutron beam. A valid approach is to use so-called correlation functions. The theory of neutron scattering from a physical system in terms of the space-time correlation function $G(\hat{r},t)$ has been developed by L. van Hove [5]. This formalism is very useful for interpreting scattering data for interacting systems (such as liquids) where the exact formulation of the scattering is extremely complicated. The pair-correlation function describes the position of nuclei in space and time, and for N nuclei it is given by:

$$G(\hat{r},t) = \frac{1}{N}\sum_{i,j}^{N}\langle\delta(\hat{r}+\hat{r}_i(0)-\hat{r}_j(t))\rangle \quad . \tag{16}$$

If $j = i$, the pair correlation function is called self pair correlation function. The space Fourier transform of the pair correlation function $G(\hat{r},t)$ is called intermediate scattering function:

$$I(\hat{Q},t) \propto \int_{-\infty}^{+\infty} G(\hat{r},t)\, exp\,(i\hat{Q}\cdot\hat{r})\, dr =$$

$$= \frac{1}{N}\sum_{i,j}^{N}\langle exp(-i\hat{Q}\cdot\hat{r}_i(0))\, exp(i\hat{Q}\cdot\hat{r}_j(t))\rangle \ . \tag{17}$$

The time Fourier transform of the intermediate scattering function leads to the so-called dynamic structure factor:

$$S(\hat{Q},\omega) \propto \int_{-\infty}^{+\infty} I(\hat{Q},t)\, exp\,(-i\omega t)\, dt =$$

$$= \int_{-\infty}^{+\infty}\int_{-\infty}^{+\infty} G(\hat{r},t)\, exp\,(i\hat{Q}\cdot\hat{r})\, dr\, exp\,(-i\omega t)\, dt \ . \tag{18}$$

Each of the functions $G(\hat{r},t)$, $S(\hat{Q},\omega)$, $I(\hat{Q},t)$ can equally describe the properties of the observed sample; however, only $S(\hat{Q},\omega)$ and $I(\hat{Q},t)$ are the quantities that can be directly accessed by neutron scattering. Eqs. (17) and (18) then relate the quantities measured by neutron scattering ($S(\hat{Q},\omega)$ and $I(\hat{Q},t)$) with the microscopic description of the sample in terms of the probability for nuclei in the sample to be found in a given position at a given time ($G(\hat{r},t)$). It is evident from Eq. (17) that $I(\hat{Q},t)$ can be divided into a coherent and an incoherent part, as we did above (Eq. (14)) for the dynamic structure factor. As discussed at the end of the previous section, the neutron scattering from biological matter is often dominated by the incoherent signal. Indeed, about 50% of the atoms in biological matter are composed of hydrogen atoms, which usually corresponds to more than 70% of incoherent signal over the total scattering signal in amorphous (*i.e.* non crystalline) samples. In the next sections we will then focus on the incoherent components of $S(\hat{Q},\omega)$ and $I(\hat{Q},t)$.

1.3 Separation of different motions

The motion of a scattering nucleus can be described as the superposition of translational, rotational and vibrational motions. The position of the nucleus can be then separated into three terms, for translational ($\hat{r}_{tr}(t)$), rotational ($\hat{r}_{rot}(t)$) and vibrational ($\hat{r}_{vib}(t)$) motions which as a first approximation can be assumed to be dynamically decoupled:

$$\hat{r}(t) = \hat{r}_{tr}(t) + \hat{r}_{rot}(t) + \hat{r}_{vib}(t) \tag{19}$$

Subsequently the incoherent parts of the functions $I(\hat{Q},t)$ and $S(\hat{Q},\omega)$ can be written as:

$$I_{inc}(\hat{Q},t) = I^{tr}_{inc}(\hat{Q},t)\cdot I^{rot}_{inc}(\hat{Q},t)\cdot I^{vib}_{inc}(\hat{Q},t)$$

$$S_{inc}(\hat{Q},\omega) = S^{tr}_{inc}(\hat{Q},\omega)\otimes S^{rot}_{inc}(\hat{Q},\omega)\otimes S^{vib}_{inc}(\hat{Q},\omega) \tag{20}$$

where the symbol $\otimes$ denotes the convolution operation.

1.4 Elastic, quasi-elastic and inelastic scattering

Dividing the incoherent intermediate scattering function into time-independent and time-dependent parts as follows:

$$I_{inc}(\hat{Q},t) = I_{inc}(\hat{Q},\infty) + I^{inel}_{inc}(\hat{Q},t) \tag{21}$$

and taking its Fourier transform (see Eq. (18)) leads to

$$S_{inc}(\hat{Q},\omega) = I_{inc}(\hat{Q},\infty)\cdot\delta(\omega) + S^{inel}_{inc}(\hat{Q},\omega) \tag{22}$$

where $I_{inc}(\hat{Q},\infty)\cdot\delta(\omega)$ is a component purely elastic and $S^{inel}_{inc}(\hat{Q},\omega)$ accounts for the energy transfer between scattered neutrons and sample nuclei and includes the information about the microscopic dynamics of the sample that are reflected on the energy-dependence of the structure factor.

$S^{inel}_{inc}(\hat{Q},\omega)$ contains contributions at small energy exchanges that can be separated from the inelastic scattering above in the so-called quasi-elastic term. This term originates from the diffusive motions of the scattering nuclei and can be described by $S^{tr}_{inc}(\hat{Q},\omega)$ and $S^{rot}_{inc}(\hat{Q},\omega)$. The vibrational motions described by $S^{vib}_{inc}(\hat{Q},\omega)$ often produce inelastic spectra outside of the quasi-elastic region and originate from the vibrational levels of molecules. Since the vibrations and librations of the particles are often much faster (i.e. they have higher energy exchange) than the diffusive motions, it is reasonable to introduce them as an inelastic background in the quasi-elastic region. Thus the quasi-elastic incoherent scattering function can be written as:

$$S^{qe}_{inc}(\hat{Q},\omega) = exp\left(-\langle u^2\rangle Q^2\right)\left[I^{dif}_{inc}(\hat{Q},\infty)\cdot\delta(\omega) + S^{dif}_{inc}(\hat{Q},\omega) + S^{i}_{inc}(\hat{Q},\omega)\right] \tag{23}$$

where the exponential factor, known as the Debye-Waller factor, accounts for the decrease in intensity due to vibrations, $S^{i}_{inc}(\hat{Q},\omega)$ is the inelastic background in the quasi-elastic region,

$S_{inc}^{dif}(\widehat{Q},\omega)$ represents the quasi-elastic term and $I_{inc}^{dif}(\widehat{Q},\infty)$ is called elastic incoherent structure factor (EISF) and is the fraction of the total quasi-elastic intensity contained in the purely elastic peak. In the experimental data (see Fig. 2) the elastic scattering appears, instead of as an infinitely sharp line, as a peak shaped curve with an energy width corresponding to the instrumental resolution. This width defines the magnitude of the timescale over which the diffusive motions are observable. If the diffusive motions are significantly slower than the time window defined by the width of the instrumental resolution, their quasi-elastic contribution cannot be distinguished from the elastic peak. On the other hand, if the diffusive motions are much faster than the timescale determined by the instrumental resolution, the quasi-elastic component will be a very broad and flat term. If the elastic component and the wider quasi-elastic component can be separated, the EISF is a measurable quantity, evaluated from the ratio:

$$EISF(\widehat{Q}) = \frac{I^{el}(\widehat{Q})}{I^{el}(\widehat{Q})+I^{qe}(\widehat{Q})} \tag{24}$$

where $I^{el}(\widehat{Q})$ and $I^{qe}(\widehat{Q})$ are the integrated (in the ω space) intensities of the elastic and quasi-elastic components of the spectra, respectively. While the width of the quasi-elastic component can reveal information on the characteristic times of the diffusive motions, the Q-dependent EISF provides information about the nature of the diffusive motion itself.

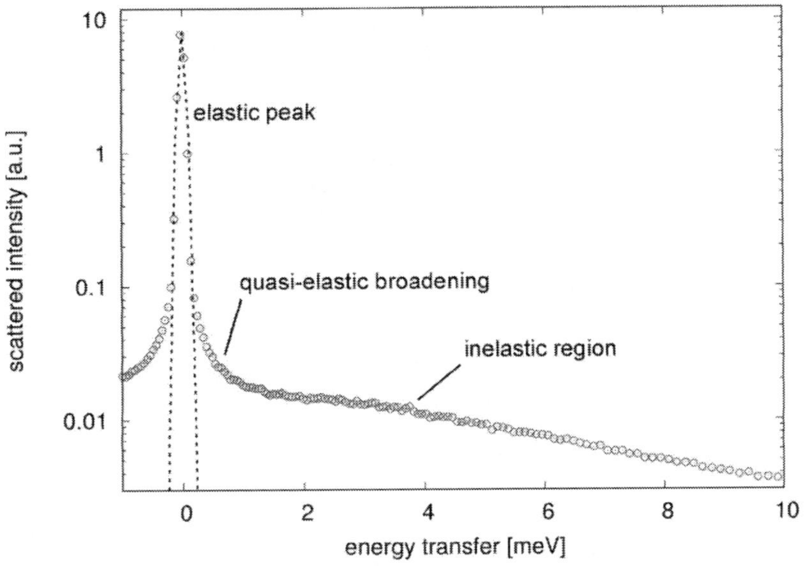

Fig. 2. Neutron scattering spectrum of a D_2O-hydrated myoglobin powder at the temperature T = 260 K measured on the time-of-flight spectrometer IN6 (Institut Laue-Langevin, Grenoble, France). The dashed line indicates the lineshape of the instrumental resolution function.

2 Spectrometers for neutron scattering

As discussed in the previous section, a neutron scattering experiment consists in determining the status of a neutron beam after interaction with the sample in terms of neutron energy and momentum, provided that the same quantities are known for the neutron beam before the interaction. Depending on the way these quantities are measured, different neutron spectroscopy techniques can be distinguished. If the change in the direction of scattered neutrons is generally defined by the position of detectors relative to the incident beam direction, the change in neutron energy can be indeed measured using different approaches, known as backscattering, time-of-flight and spin-echo. A fourth approach, called three-axis, will not be discussed here. Every technique is characterized by different energy and momentum transfers that can be measured, which in turn define the time and space scales of the structural dynamics that can be probed, as represented in an approximate way in Figure 3. The exact ranges are of course defined by the details and the technical parameters of a given spectrometer.

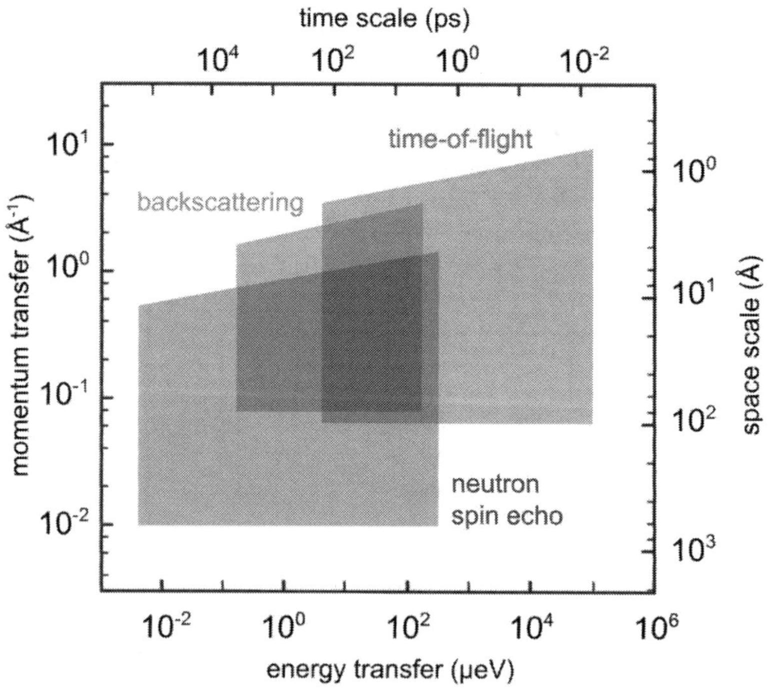

Fig. 3. The energy and momentum transfer ranges probed by the different neutron spectroscopy techniques provide access to molecular motions at different space and time scales.

2.1 Neutron backscattering spectroscopy

Neutron backscattering spectroscopy uses the principle of backscattering, which refers to the reflection of a neutron wave back to the same direction it comes from. This technique makes use of a Bragg angle of 90° to analyze and select the wavelength of neutrons after interaction with the sample, which allows to obtain a high energy resolution, even below 1 μeV. Since the Bragg angle in the analyzer cannot be changed to select different energies, different energies transfers are achieved by varying the energy of the incoming beam. This is implemented either changing the lattice constant in the monochromator crystal by changing its temperature or by Doppler effect obtained by periodically translating monochromator crystal along the reflection direction. The disadvantage of these methods for changing the energy transfer is the inherent limited energy range that can be explored. Examples of neutron backscattering spectrometers are SPHERES at the Maier-Leibnitz Zentrum in Garching (Germany) and IN16B at the Institut Laue-Langevin in Grenoble (France).

2.2 Time-of-flight spectroscopy

In the time-of-flight approach the energy of the scattered neutrons is determined by measuring their kinetic energy after interaction with the sample. The kinetic energy is calculated using the measurement of the time neutrons take to travel along the distance between the sample and the detector. The same principle can be used for selecting the energy of the neutron beam impinging on the sample, by means of a system of synchronized rotating choppers. Examples of time-of-flight spectrometers are TOFTOF at the Maier-Leibnitz Zentrum in Garching (Germany) and IN5 at the Institut Laue-Langevin in Grenoble (France).

2.3 Neutron spin-echo spectroscopy

Neutron spin-echo spectroscopy is the only neutron technique that enables accessing the dynamics of nuclei in the sample directly in the time domain. The main difference with respect to the other methods described above is that neutron spin-echo provides a direct estimation of the energy exchange, instead of calculating it by subtracting the neutron energy after interaction with the sample to energy before the interaction. The main idea at the basis of neutron spin-echo is to use Larmor precession of polarized neutrons travelling inside a magnetic field as an internal clock. Neutrons travel in identical magnetic fields before and after interacting with the sample and the spin orientation is inverted at the sample position. In elastic conditions, neutron polarization is restored at the outcome of the second magnetic field, otherwise, if energy has been exchanged with the sample, a change in the polarization will be measured, from which the inelastic scattering can be calculated in terms of the intermediate scattering function $I(\widehat{Q},t)$. Examples of neutron spin-echo spectrometers are the NGA spectrometer at the NIST Center for Neutron Research (US) and IN15 at the Institut Laue-Langevin in Grenoble (France).

3 Neutron scattering probes biomolecular dynamics

In this section some applications of neutron scattering to investigate the dynamical behavior of biological matter are reviewed. Depending on the physical state of the sample and on the space-time scale probed by a given neutron spectrometer, the complexity of the structural dynamics contributing to the neutron scattering signal makes the data interpretation more or less challenging. In the case of proteins, for example, assuming that the $H_2O \rightarrow D_2O$ exchange in the buffer minimizes the solvent contribution to the scattering signal, the remaining signal arises from different classes of molecular motions (see Fig. 4). Local fluctuations at the side-chains and backbone level characterize the dynamics in the ps-ns time scale; internal dynamics are also composed by relative motions of protein domains, which can occur from ps up to μs; finally, roto-translational diffusive motions of the entire macromolecules can be detected in the ns up to ms time scale, depending on the space scale probed by the neutron spectrometer and on the molecular weight of the macromolecule. The use of hydrated protein powders, where all motions but the internal dynamics are essentially suppressed, is a way to simplify the data interpretation and focus only on internal dynamics.

In this section three different examples have been selected that illustrate the use of elastic, quasi-elastic and inelastic neutron scattering for studying the behaviour of biomolecules in terms of dynamical processes at a molecular level:
1) the role of hydration water in triggering the onset of functional motions in soluble proteins [6] and how this can be implicated in protein pathological states [7];
2) the collective dynamics of lipid membranes as a function of temperature [8];
3) the diffusive dynamics of proteins in crowded conditions mimicking the cellular context [9,10].

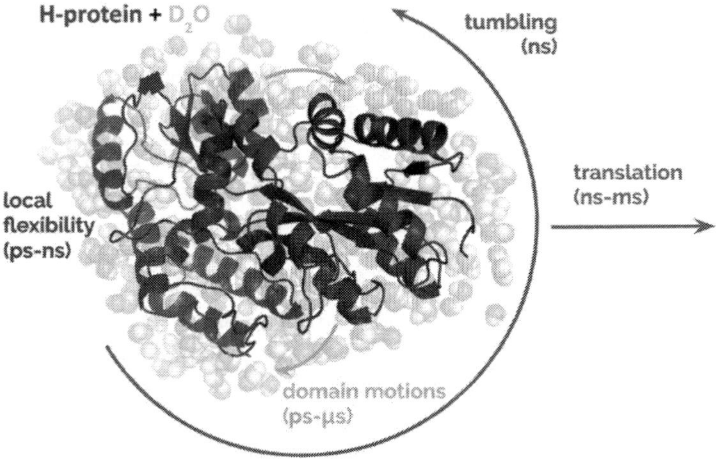

Fig. 4. Different classes of molecular motions characterize biological macromolecules: local side-chains and backbone fluctuations occurring in the ps-ns time scale, internal domain motions (ps-µs) and roto-translational motions of the entire molecule, occurring in the ns up to ms time scale, depending on the probed space scale and on the size of the macromolecule.

3.1 Quasi-elastic neutron scattering reveals the role of hydration water in protein dynamics

Biological macromolecules such as soluble proteins are surrounded by a layer of water molecules (the so-called hydration water) interacting with the macromolecule surface. The active role of hydration water in macromolecular functions such as enzyme reactions, molecular recognition, allostery, and charge transfer is now largely recognized. In most completely dry proteins biological activity is lost, while it is recovered when the first hydration layer is present, thus making a protein together with its first hydration layer the biologically active entity. The interaction between protein and water molecules is essentially of electrostatic nature, with an extended hydrogen-bonded network connecting protein and water. The fluctuations of such a network are accompanied by breakage and formation of water-protein hydrogen bonds and allow for functionally important protein motions. The nature of the coupling between water and protein dynamics is a matter of extensive debate [11]. A way to tease apart the complex ensemble of water-protein motions is to extend experiments down to cryo-temperatures. When lowering the temperature below ~240 K hydrated proteins undergo a transition where the large-amplitude anharmonic motions on the picosecond timescale, which are related to functional motions, are frozen. The origin of this transition, which has been named protein dynamical transition [12] and is paralleled by a transition in the picosecond motions of hydration water at a similar temperature [13], is controversially debated in the biophysical community [11]. However, there is a general agreement about the primary role played by hydration water dynamics in both protein dynamics and function. Indeed, when the hydration level is lowered below monolayer coverage both the dynamical transition and the biological activity of a protein are suppressed [14].

As discussed above, the incoherent scattering cross-section of a hydrogen atom is larger by about two orders of magnitude than that of all other atoms in a protein (see Fig. 1), including its isotope deuterium, and thus dominates the neutron scattering signal in non-crystalline protein samples. Replacing all hydrogen atoms in a protein by deuterium (a procedure called perdeuteration) is then a way to enhance the contribution from hydration water dynamics to the incoherent scattering signal and minimize that from protein dynamics. In perdeuterated proteins put in the form of freeze-dried powder hydrated at a level of ~0.4 gram of H_2O per gram of protein, corresponding roughly to a monolayer coverage of the protein surface, more than 70% of the incoherent neutron scattering signal comes from hydration water.

In the work reviewed in this section [6] hydration water dynamics has been studied on the surface of the globular Maltose Binding Protein (MBP) and of the intrinsically disordered human protein tau. By combining protein perdeuteration and quasi-elastic neutron scattering, evidence has been provided that the onset of translational water motions on tau and MBP occurs at the same temperature where the proteins undergo the dynamical transition. The diffusive dynamics of a water molecule can be described, in the time scale probed by neutron scattering, as a translational motion of its centre-of-mass and a rotation around the centre-of-mass. The model used to analyze experimental data assumes that water molecules either translate, rotate or remain immobile. Indeed, the fitting quality has been proved to improve with a model function where the translational and rotational terms are added rather than convoluted (χ^2 at least 20% higher on average, but even 50% higher at temperatures above 250 K). This can be explained by the fact that within the ns-ps time scale probed by the neutron spectrometer two populations of water molecules exist with different dynamical properties: one formed by water molecules strongly interacting with the protein surface and/or inside protein docking sites (represented by the rotational term) and another one formed by less hindered water molecules whose rotational motion is faster than the time window accessible by the neutron spectrometer (represented by the translational term). Such heterogeneity of water motions in the first hydration shell has been also observed by other techniques and on different time scales [15].

A novelty of the work reviewed here [6] was also the global fitting approach adopted, in which the distance between neutron experimental data and the model function (i.e. the χ^2 function) was minimized in the two-dimensional space defined by momentum transfer and energy transfer. Compared to the fitting procedure usually adopted in the literature (i.e. fitting the experimental data in the energy transfer domain at every momentum transfer value and then analyzing the momentum transfer dependence of the fitting parameters), the approach proposed in [6] allows from one hand to minimize the number of free parameters and from the other hand to identify dynamical contributions to the quasi-elastic spectrum arising from confined motions (such as rotational motions). These contributions have indeed a typical length scale (like, in the present case, the radius of a rotational motion defined by the H-O distance in the water molecule) so that their intensity depends on the momentum transfer and can vary significantly in its accessible range. From the global fit analysis, for each temperature point fractions of the total scattering intensity originating from water molecules that undergo translational or rotational diffusion, or that remain immobile have been then extracted. The plots in Fig. 5 show these fractions as a function of temperature for both tau and MBP. The translational fraction displays a steep increase above ~240 K, while the rotational one starts increasing at ~200 K and reaches a plateau at ~250 K (see Fig. 5). Up to ~200 K, nearly all water molecules are detected as immobile on the timescale determined by the energy resolution of the neutron spectrometer. With the resolution used in this work a molecule is defined as "immobile" when its motion is slower than ~2 ns. A full population of such "immobile" water molecules up to ~200 K is in agreement with the presence of a calorimetric glass transition of hydration water at ~170 K, as actually measured in hydrated powders of other globular proteins [16].

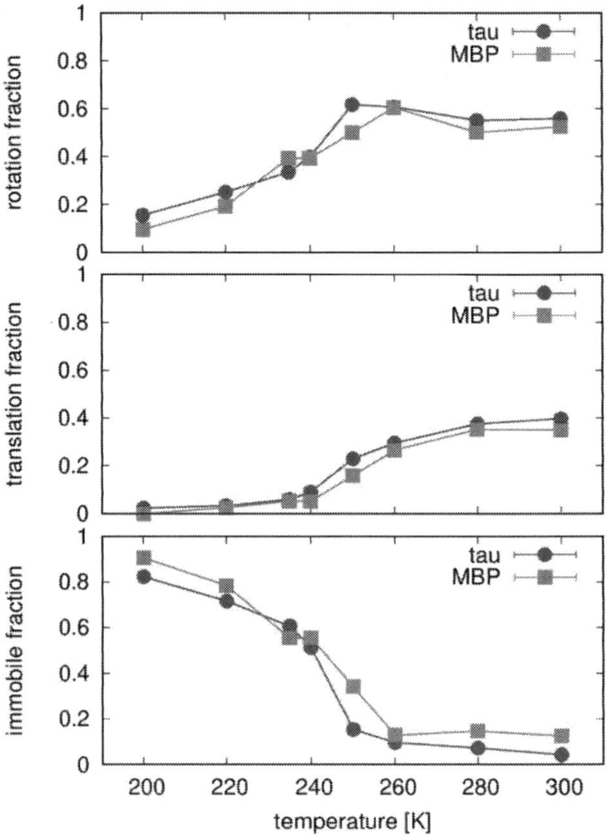

Fig. 5. Fractions of the different components of water dynamics in both tau (purple circles) and MBP (green squares) extracted from the analysis of quasi-elastic spectra as a function of temperature: water molecules not moving in the dynamic window investigated (lower panel), centre-of-mass translation of water molecules (mid panel) and rotation of water molecules around their centre-of-mass (upper panel). An onset of translational motions is evident at ~240 K, while rotational motions are present already at 200 K. Data extracted from [6].

The experimental work reviewed in the present section proved the notion already proposed by molecular dynamics simulations [17,18] that the protein dynamical transition detected by neutron scattering at ~240 K [12] is connected to the onset of water translational diffusion via relaxation of the protein-water hydrogen bond network. Molecular dynamics simulations on the same systems studied by neutron scattering were also used to confirm the experimental results in [6], but this computational part will not be reviewed here. An essential role in the present approach was played by the line width analysis of quasi-elastic data that provided insight into the nature of motions of water molecules in the hydration layer not provided by elastic neutron scattering [13]. The fact that both the globular protein MBP and the intrinsically disordered protein tau display the same mechanism suggested, unexpectedly, that this connection is independent of the degree of order in the protein conformation, and is a general behavior of soluble proteins.

The same approach by protein perdeuteration and neutron scattering was applied to the study of the dynamics of human tau when its amyloid pathological form is induced [7]. The biological function of human tau is the regulation of microtubule stability in neurons. In the case of Alzheimer disease, tau aggregates and forms so-called paired helical filaments (PHF) [19]. Quasi-elastic neutron scattering spectra were measured on both native and aggregated deuterated tau at 280 K. Aggregated tau displays a larger quasi-elastic broadening, in qualitative agreement with the enhanced dynamics of hydration water evidenced in the same sample by elastic neutron scattering. Quasi-elastic data were fitted with the same approach described above [6]. The fraction of water molecules undergoing translational diffusion around the tau PHF was found to be ~25% higher than around native tau. The translational diffusion coefficient and the rotational rate were ~11% and ~17% higher, respectively, for the PHF hydration water. These results provided experimental evidence that hydration water mobility of tau increases upon amyloid aggregation. Such an increase of water dynamics on the surface of tau PHF suggested a scenario in which the hydration water mobility favors the formation of tau amyloid fibers by providing entropic compensation to the disorder-order (from the intrinsically disordered conformation to the PHF alignment) transition induced in the protein during PHF growth.

3.2 Inelastic neutron scattering probes collective dynamics of lipid membranes

Understanding the physical properties of lipid membranes is one of the main objectives of biophysical sciences. The use of model systems for lipid bilayers is aimed at obtaining information on the behavior of cell membranes in physiological conditions. In particular, the structural organisation of membranes is supposed to determine their collective response to cellular interactions. The collective dynamics of lipid molecules forming a membrane affect significantly its physical properties. In the cellular context, collective motions are involved in different biological functions. As an example, the transport of small molecules through the membrane needs correlated molecular motions of the lipid acyl chains and the corresponding density fluctuations in the membrane plane [20]. This kind of short range collective motions can be studied by inelastic neutron scattering. One of the first inelastic studies, of which some results are reviewed in this section, focused on the collective dynamics of the lipid acyl chains in the model system 1,2-dimyristoyl-sn-glycero-3-phoshatidylcholine (DMPC)-d54 [8]. Using selective deuteration of the chains to enhance the respective motions over other contributions to the inelastic scattering cross section, the dynamic structure factor $S(Q,\omega)$ was measured in both the gel and the fluid phase of the DMPC bilayer. The temperature dependence of the excitations in the dispersion minimum was also investigated in the vicinity of the main phase transition, occurring at T_c = 21.5° C in the deuterated compound. The experiments were performed at the spectrometers IN12 and IN3 at the Institut Laue-Langevin (ILL) in Grenoble, France. The probe energy was of the same order as the excitations (meV) and the energy resolution ~300 μeV. The direction of the scattering vector $\widehat{Q}$ was in the membrane plane in order to use the same experimental set-up to measure both the static structure factor $S(Q_r)$, where Q_r is the lateral momentum transfer, and the dynamic structure factor $S(Q_r, \omega)$.

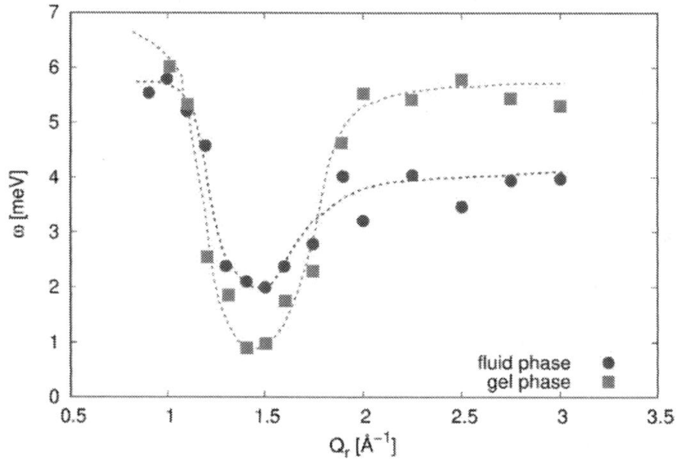

Fig. 6. Dispersion relations in both the fluid (T = 30° C, purple circles) and the gel phase (T = 20° C, green squares) of the DMPC bilayer. The energy at the minimum is lower for the gel phase than for the fluid phase. Data extracted from [8].

Using the energy positions of the Brillouin peaks in the inelastic scans at different Q values in the range $0.7 < Q < 3.0$ Å^{-1} the dispersion relation in both the gel and the fluid phase was determined (Fig. 6). The results in Fig. 6 show that the energy at the minimum of the dispersion relation is actually deeper in the gel than in the fluid phase. From the biological point of view, determining the dispersion relation in both phases can help understanding how molecules can be transported across to as well as along the plane of the lipid bilayer, in particular when considering that the collective dynamics at the level of the lipid tails may couple with the self-diffusive dynamics of the lipid molecules. Free area enabling diffusive dynamics can indeed be induced by the collective density waves detected in [8], with the wave frequency providing the vibrational bath for thermally activated conformations of the lipid chains which enable the formation of free volume.

3.3 High-resolution neutron scattering simultaneously probes internal and diffusive protein dynamics in crowded conditions

Biological molecules *in vivo* (i.e. in the cellular environment) are characterized by conditions of crowding and geometrical confinement [21]. Protein and RNA concentration inside the cell is of the order of ~300-400 mg/ml, corresponding to about 30 % of the cytoplasm volume. For example, the concentration of hemoglobin alone is ~350 mg/ml inside the red blood cells. Several works (see e.g [22] and references therein) showed that chemical equilibria and chemical reactivity inside the cell are perturbed by non-specific solute-solute interactions and excluded volume effects. However, studies on the dynamical properties of proteins in crowded conditions are still not very abundant. In the last years quasi-elastic neutron backscattering at the spectrometers IN16 and IN10 (ILL, Grenoble) has been used to obtain extensive experimental data on protein self-diffusion on the nanosecond time and nanometer length scales in crowded aqueous solutions of bovine serum albumin (BSA) [9], β-lactoglobulin (BLG) [10] and other proteins at high concentration [23].

In the first work on BSA [9], quasi-elastic spectra were modeled by a convolution of two lorentzian terms, one accounting for fast internal and interdomain motions within the protein and the other one for the convolution of the translational and rotational diffusion of the entire protein. From the Q-dependence of the linewidth of the second term the total diffusion coefficient D was determined in the low concentration extrapolation and compared with the same parameter obtained by dynamic light scattering to validate the approach. An analytical method was then developed to extract the translational diffusion coefficient D_t from D also for the high concentration regime. The study of D_t as a function of the protein volume fraction showed that crowding has a strong effect on the protein self-diffusion already on the nanoseconds time scale, the diffusion coefficient at high protein concentration being strongly decreased compared to the dilute limit. Such a slowing-down effect on the protein diffusion was found to be mainly caused by hydrodynamic interactions since it could be described with very good accuracy in terms of colloidal short-time self-diffusion.

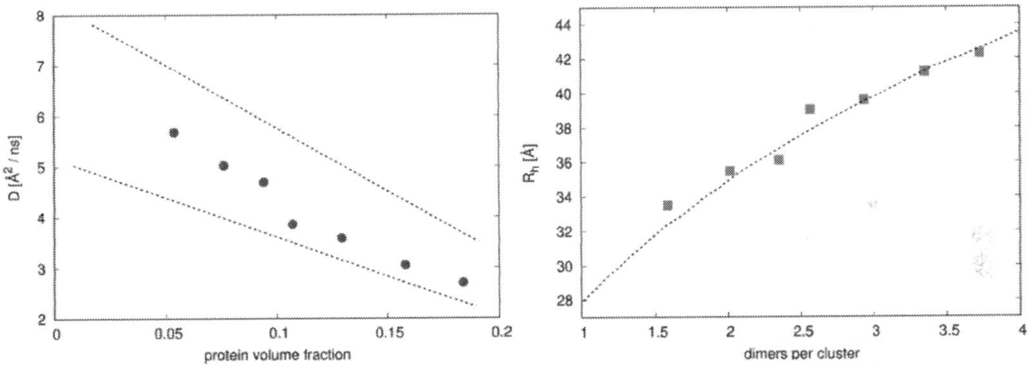

Fig. 7. Left panel: apparent diffusion coefficients D (purple circles) of BLG obtained by the analysis of neutron backscattering data as a function of the protein volume fraction. The black dotted lines are calculated considering the diffusion coefficient of a BLG dimer (upper line) and a cluster of four BLG dimers (lower line). Right panel: effective hydrodynamic radii R_h (green squares) of the BLG clusters as a function of the number of dimers per cluster calculated from small angle X-ray scattering data. The black dotted line is a fitting curve with a function $R_h \propto N^a$ yielding $a = 0.32 \pm 0.02$. Data extracted from [10].

The second work reviewed in this section [10] used an analogous approach to focus on the clustering of BLG dimers in crowded conditions. The phenomenon of clustering in protein aqueous solutions is interesting from both a fundamental and a practical point of view. The mechanisms inducing cluster formation have been investigated in model systems, yet their understanding may help to control processes like protein self-assembly, drug delivery, protein aggregation and crystallization. In BLG solutions clustering can be induced just by increasing the protein volume fraction, without the necessity for changing other chemical parameters like salt concentration or pH. In physiological conditions and at low volume fraction BLG is found predominantly as a dimer.

A detailed description of cluster formation was obtained from the self-diffusion coefficients measured by neutron backscattering as a function of protein volume fraction in the range 0.05 - 0.2 (Fig. 7 left panel). In the high concentration regime, the apparent diffusion coefficients were not in agreement with those calculated assuming the presence of BLG dimers while a better agreement was obtained with BLG clusters formed by four dimers (dotted lines in Fig. 7 left panel). More precise information on the average hydrodynamic size of the clusters was derived from a fitting of the apparent diffusion coefficients D at each different protein concentration using the effective hydrodynamic radius R_h as only free parameter (Fig. 7 right panel). A scaling law $R_h \propto N^a$ was used to fit the data (Fig. 7 right panel), with the fitting exponent found to be $a = 0.32 \pm 0.02$. An exponent ~1/3 is consistent with the presence of quite compact clusters of BLG dimers whose size increases with increasing protein concentration. The combination of backscattering results with complementary data from small angle X-ray scattering and neutron spin echo (not described here) enabled the authors to obtain a robust description of the presence of clusters in BLG aqueous solutions that are static on the nanosecond time scale accessed by neutron spectroscopy. The analysis proposed in [10] can be used as a general approach to provide a quantitative description of the structure and the diffusive dynamics of macromolecular assemblies at high concentration.

References

1. S. W. Lovesey, *Theory of Neutron Scattering from Condensed Matter* (Clarendon Press, Oxford, 1984)
2. G. L. Squires, *Introduction to the Theory of Thermal Neutron Scattering* (Courier Corporation, 1996)
3. M. Bée, *Quasielastic Neutron Scattering* (Adam Hilger, 1988)
4. E. Johnson and L. Robinson, Rev. Sci. Instrum. **60**, 3447 (1989)
5. L. Van Hove, Phys. Rev. **95**, 249 (1954)
6. G. Schirò, Y. Fichou, F.-X. Gallat, K. Wood, F. Gabel, M. Moulin, M. Härtlein, M. Heyden, J.-P. Colletier, A. Orecchini, A. Paciaroni, J. Wuttke, D. J. Tobias, and M. Weik, Nat. Commun. **6**, 6490 (2015)
7. Y. Fichou, G. Schirò, F.-X. Gallat, C. Laguri, M. Moulin, J. Combet, M. Zamponi, M. Härtlein, C. Picart, E. Mossou, H. Lortat-Jacob, J.-P. Colletier, D. J. Tobias, and M. Weik, Proc. Natl. Acad. Sci. U. S. A. **112**, 6365 (2015)
8. M. C. Rheinstädter, C. Ollinger, G. Fragneto, F. Demmel, and T. Salditt, Phys. Rev. Lett. **93**, 108107 (2004)
9. F. Roosen-Runge, M. Hennig, F. Zhang, R. M. J. Jacobs, M. Sztucki, H. Schober, T. Seydel, and F. Schreiber, Proc. Natl. Acad. Sci. U. S. A. **108**, 11815 (2011)
10. M. K. Braun, M. Grimaldo, F. Roosen-Runge, I. Hoffmann, O. Czakkel, M. Sztucki, F. Zhang, F. Schreiber, and T. Seydel, The Journal of Physical Chemistry Letters **8**, 2590 (2017)
11. G. Schirò and M. Weik, Journal of Physics: Condensed Matter **31**, 463002 (2019)
12. W. Doster, S. Cusack, and W. Petry, Nature **337**, 754 (1989)
13. K. Wood, A. Frölich, A. Paciaroni, M. Moulin, M. Härtlein, G. Zaccai, D. J. Tobias, and M. Weik, J. Am. Chem. Soc. **130**, 4586 (2008)
14. T. Kleinert, W. Doster, F. Post, and M. Settles, in *Conference Proceedings-Italian Physical Society* (Editrice Compositori, 1992), pp. 127–127

15. S. K. Pal, J. Peon, and A. H. Zewail, Proc. Natl. Acad. Sci. U. S. A. **99**, 1763 (2002)
16. G. Schirò, M. Fomina, and A. Cupane, J. Chem. Phys. **139**, 121102 (2013)
17. M. Tarek and D. J. Tobias, Phys. Rev. Lett. **88**, 138101 (2002)
18. A. L. Tournier, J. Xu, and J. C. Smith, Biophys. J. **85**, 1871 (2003)
19. K. S. Kosik, C. L. Joachim, and D. J. Selkoe, Proc. Natl. Acad. Sci. U. S. A. **83**, 4044 (1986)
20. S. Paula, A. G. Volkov, A. N. Van Hoek, T. H. Haines, and D. W. Deamer, Biophys. J. **70**, 339 (1996)
21. R. J. Ellis, R. John Ellis, and A. P. Minton, Nature **425**, 27 (2003)
22. A. P. Minton, Journal of Biological Chemistry **276**, 10577 (2001)
23. M. Grimaldo, F. Roosen-Runge, F. Zhang, F. Schreiber, and T. Seydel, Quarterly Reviews of Biophysics **52**, (2019)

AUTHOR INDEX

Bhatt Jayesh S. 86
Combet Sophie 1
Coquelle Nicolas 19
Cousin Fabrice 101
Fadda Giulia 101
Franzetti Bruno 76
Gabel Frank 76
Gerelli Yuri 116
Härtlein Michael 76
Ibrahim Ziad 76
Jeffries Cy M. 41
Mahieu Emilie 76
Martel Anne 76
Moulin Martine 76
Pietras Zuzanna 41
Schirò Giorgio 134
Svergun Dmitri I. 41
Zaccai Nathan Richard 19